Gifted Earth

Gifted Earth

The Ethnobotany of the Quinault and Neighboring Tribes

DOUGLAS DEUR

AND THE KNOWLEDGE-HOLDERS OF

THE QUINAULT INDIAN NATION

Oregon State University Press *Corvallis*

published in cooperation with the
Quinault Indian Nation

This material is based upon work assisted by a grant from the Department of Interior, National Park Service. Any opinions, findings, and conclusions or recommendations expressed in this material are those of the author(s) and do not necessarily reflect the views of the Department of the Interior.

The John and Shirley Byrne Fund for Books on Nature and the Environment provides generous support that helps make publication of this and other Oregon State University Press books possible.

Library of Congress Cataloging-in-Publication Data
Names: Deur, Douglas - author.
Title: Gifted earth : the ethnobotany of the Quinault and neighboring tribes / Douglas Deur and the Knowledge-Holders of the Quinault Indian Nation.
Description: Corvallis : Oregon State University Press, [2019] | Includes bibliographical references and index.
Identifiers: LCCN 2019017059 | ISBN 9780870719653 (original trade pbk. : alk. paper)
Subjects: LCSH: Quinault Indians—Ethnobotany—Washington (State)—Olympic Peninsula. | Indians of North America—Ethnobotany—Washington (State)—Olympic Peninsula.
Classification: LCC E99.Q6 D48 2019 | DDC 979.7004/9794—dc23
LC record available at https://lccn.loc.gov/2019017059

∞ This paper meets the requirements of ANSI/NISO Z39.48-1992 (Permanence of Paper).

First published in 2022 by Oregon State University Press
Second printing 2024
Printed in the United States of America

Oregon State University Press
121 The Valley Library
Corvallis OR 97331

Published in cooperation with
Quinault Indian Nation
1214 Aalis Drive
Taholah, WA 98587

Front cover: Douglas fir illustration from: P. Mouillefert, 1892–1898. *Traité des arbres et arbrissaux, Atlas* (top); "Quinault War Canoe" from E. S. Curtis, 1913, *The North American Indian, Volume* 9 (bottom).

Back cover: Salmonberry, from E. H. Harriman and C. H. Merriam, 1901, *Harriman Alaska Expedition*.

Frontispiece: Watching the canoes arrive at Taholah in traditional cedar bark clothing, "Paddle to Quinault" Canoe Journey 2013.

Contents

Sharing: A Prologue

In any publication containing Native American knowledge, an urgent tension exists between the need for privacy and the need for disclosure.

On one hand, sharing cultural information with non-Native readers is inherently problematic. Sharing with a general audience can introduce new threats: the loss of a tribe's cultural distinctiveness; increased competition for scarce resources; the disrespectful or even dangerous misuse of cultural knowledge. On the other hand, not sharing also carries potential costs. Indeed, withholding information can contribute to the degradation of an entire culture: essential knowledge might not find its way to tribal youth, to friends and allies, or to those non- Native neighbors who potentially affect the integrity of tribal cultures and homelands through their countless acts, official or everyday. If non-Native people do not comprehend and value those things that matter to Native people—such as plants, special places, or the basic cultural practices linked to them—there is a genuine risk that those things will be destroyed inadvertently, and without much notice or concern from the non-Native world. Done respectfully, disclosure can help foster broad understanding and empathy, and buttress protections for many of the things tribes value most—the very things that help hold together Native cultures and communities.

Each time a tribe considers sharing cultural knowledge in a book or some other venue, knowledge-holders consider the balance between the costs and the benefits of disclosure. This book is no exception. Ultimately, the Quinault Indian Nation determined that sharing the information in this book was in everyone's best interest. Sharing, the tribe concludes, is the right choice for this material, in spite of the potential risks of disclosure. It is from that perspective that this book is written.

As the pages that follow attest, many people were involved with the development and design of this book—tribal staff, elders, and many others. The names of many of these contributors appear throughout the pages that follow. The entire book, and the research it encapsulates, also reflects the guidance of Quinault cultural expert Justine E. James Jr. For many years, Justine has served as the cultural resource specialist in the Quinault Division of Natural Resources and worked as the lead cultural officer for his tribe. His father, Justine E. James Sr., was among the principal interviewees for this project, sharing his rich memories of plant gathering, fishing, and hunting on the reservation, but also in many other parts of the Quinault homeland—especially on lands now within Olympic National Park. For many years, in the course of his work for the tribe, Justine James Jr. had been asked to officially characterize the cultural effects of industrial logging and development in places within the Quinault homeland. He long recognized that those requests almost exclusively focused on archaeological sites and seldom included other essential categories—including places and natural resources, such as culturally significant plant communities, that are of importance to the living culture. Some of these plant-gathering sites have been such important venues for subsistence and cultural activities across generations of tribal members that they represent premier "cultural sites," known to almost every member of the tribe—yet, they have often remained poorly documented, excluded from the discussion of damaging effects, and thus vulnerable to potential destruction and despoliation.

In response, working closely with other Quinault Division of Natural Resources staff, Justine began to develop a standardized list of culturally important plants, along with their Quinault names and basic information about their cultural significance and uses. This list he shared with key people, such as natural resource planners, who might otherwise ignore these culturally important plants—that they might begin to understand, that they might begin to see the value and richness of Quinault's ethnobotanical heritage, that they might help protect plant communities that continue to sustain the tribe. Sometimes it worked. Beyond this protective value of his work, he saw that tribal members too were very eager for written information on the unique values of plants in the food, medicinal, material, and spiritual dimensions of their culture. Based on that initial experience, Justine saw the need for a more thorough document encapsulating Quinault ethnobotanical information. That is where our formal collaborations began.

With Justine as catalyst, the Quinault Indian Nation initially recruited me to research and develop ethnobotanical guidebooks for a very specific purpose: the preservation of traditional ethnobotanical knowledge for the benefit of tribal members and employees. The book you see now has been derived largely from original in-house guidebooks developed for that use. These in-house Quinault Indian Nation guidebooks presented plant information in an easy-to-use format similar to other plant field guides, but with information obtained solely from the elders and plant users of the Quinault Indian Nation. The guidebooks developed for the tribe as part of our initial work now appear in classrooms for tribal youth, in the health and dietary programs of Quinault Indian Nation, and on the shelves of natural resource managers who work for the tribe, so that they might factor the cultural value of plants into their larger mission.

With this book, the Quinault Indian Nation has elected to share some of this information, so that non-Native people might understand the enduring sophistication and vitality of Quinault practices of plant use. The tribe chose to share this information so that their neighbors who are hungry or in need of healing may benefit. The tribe chose to share this information so that people might appreciate that—in spite of their reservation and treaty rights—the Quinault people have lost access to some of their most important plant-gathering areas. The freedom to access and manage places of long-standing importance to the tribe has thus diminished—in the carefully managed mountain huckleberry patches now within Olympic National Park, on coastal prairie clearings, along the increasingly developed shoreline of Grays Harbor.

As restricted access hurts the culture, elders note, it also undermines the integrity of the plant communities that have been linked to that culture over deep time. Perhaps most importantly, the tribe chooses to share this information so that the many people who now live in or affect Quinault homeland might begin to see the value of this region's rich botanical heritage, to respect rather than to ignore, to protect rather than to destroy.

We also share this information with the understanding that the plants we describe are relatively ubiquitous across the Pacific Northwest coastline—and that, by learning about the plants valued by Quinault families, readers will gain a deeper familiarity and respect for the plants in their own backyard. In writing this book, we are not suggesting that Quinault country is a uniquely suitable plant gathering venue for everyone. The Quinault Indian Reservation is reserved for tribal use only and is governed by the sovereign Quinault Indian Nation. Those who enter are required to follow all laws and rules of the tribe. This is consistent with the role of the tribe as hosts on their own land, and is in line with countless generations of teachings about how we show proper respect to one another. The plants mentioned in this book used to flourish throughout Western Washington but are now a very limited resource and should be treated with care. It is the duty of all humanity to understand what we have in order to preserve these precious resources for future generations.

The information in this book is shared openly with the larger world, but the contents remain the intellectual property of the Quinault people and the individual Quinault contributors mentioned by name. Copyright in this book is held by the Quinault Indian Nation. A portion of the proceeds from sales of this book go to the tribe as royalties, and OSU Press will provide free and discounted copies to the tribe to support the transmission of plant-use knowledge and traditions for generations yet to come. This is all for the greater good, and by design. It is our sincere hope that the non-Native world gains even more respect for the plants and the culture of the Quinault and that, in turn, this respect contributes to the well-being of both people and plant habitats in Quinault country. If this book helps support that goal, even in a small way, it will be worthy of the time, the energy, and the support of the Quinault knowledge-holders who contributed to its pages.

Douglas Deur, Oregon Coast, 2021

Acknowledgments

This book was possible only through the work of many people who shared and supported its goals. Justine E. James Jr. saw the need for this information to be recorded and carried forward in time; he first envisioned and continuously directed the study on which this guidebook is based. Many members of the Quinault Indian Nation staff also played key roles in supporting the work, including but not limited to Janet Clark, Leilani "Lani" Chubby, Christina Breault, Bennett Hestmark, Roberta Harrison, Zeke Serrano, Larry Workman, Jess Helsley, Roseann Reed, Caroline Martorano, Daniel Ravenel, Guy Capoeman, and the members of the Quinault language team. Early phases of the research were supported by Quinault research interns, including Brenda Rhoades and Michael McCoy. Other phases of the study were supported by other technical experts: Woodland Hood provided extensive research support at several stages of the project; Steve Mark aided in certain archival tasks; Dr. Nancy Turner (University of Victoria) and Carla Cole (National Park Service) provided important peer review and revision; Susan Campbell and Dr. Tricia Gates Brown edited the manuscript; Mary Hyde oversaw the graphic design of the original Quinault tribal guidebooks, and her work has much influenced the look and style of the present book. A number of friends and colleagues—some Quinault and some not—provided photos and are identified in the photo credits. Many others helped in important but smaller ways. The staff of OSU Press also provided their cheerful, wise counsel and all manner of technical help in the production of the present book. I offer my deep gratitude, and apologies, to anyone who has assisted in bringing this book to fruition but is not mentioned here by name.

It is to the Native American knowledge-holders, however—those who agreed to be interviewed and contribute to this project—that this book is dedicated: Katherine Barr, Richard Sivonen, Conrad Edwards, Justine James Sr., Justine James Jr., Harvest Moon, Francis Rosander, Tekie Rosander, Micah Masten, Marlene Johnson, Guy Capoeman, David Martin, Clifford "Soup" Corwin, Margaret Payne, Vernon Lorton, Phillip Antone Hawks, Chris Morganroth, Mel Moon, Leta Shale, Dennis Martin, Lucille Quilt, Alicia Figg, Phil Martin Sr., Eugena Hobucket, Jared Eison, Conrad Williams, Celina Charles, Mandy Hudson-Howard, Marlene Hanson, Glenna Gardner, Lloyd Smith, and Gerald Ellis. They are principally Quinault Indian Nation tribal members, though their affiliations also include the Quileute Tribe, Quinault enrollees affiliated with Chinook Indian Nation, and others. This book would not exist if not for their significant contributions. We also thank the organizations that provided financial and other support in producing this book: the National Park Service, the Institute of Museum and Library Services, the US Department of Agriculture Wetlands Program, and the Olympic Peninsula Intertribal Cultural Advisory Committee.

This book also would not have been possible without the countless generations of Quinault ancestors, carrying forward plant knowledge and sharing it with the youth of their time. They are our contributors' teachers, and their teachers before them. Their example has provided inspiration at every stage of this project.

Interviewing and research will continue after the production of this book. We offer sincere thanks to all who will continue to expand our understanding of Quinault plant traditions into the future. We also offer thanks to the plants and plant communities, who continue to hold up their end of our ancient interspecific bargain.

This Ethnobotanical Guidebook: How It Came to Be and How to Use It

The ethnobotanical traditions of the Quinault have ancient roots and continue to sustain tribal members into the present day. This guidebook provides a small window into this rich living tradition, based on knowledge shared by Quinault knowledge-holders, past and present. They have shared this information both to celebrate the richness of their plant traditions and to help sustain the traditions.

This guidebook represents the outcome of a multiyear study, directed by Dr. Douglas Deur and guided by the vision of Justine E. James Jr., of the Quinault Indian Nation and Division of Natural Resources. Tribal members had long expressed the view that past ethnobotanical writings regarding their ancestral tribes were very thin, often inaccurate, and insufficient for carrying forward the plant knowledge they treasure so much. The few available published accounts of Quinault plant use had stark limitations. The most focused treatment, in Erna Gunther's 1945 book, *The Ethnobotany of Western Washington,* had been informed by only a few brief conversations with Quinault people whom Gunther had encountered off-reservation, and is striking for its inaccuracies and oversights relating to the Quinault. In the one classic ethnographic study of the Quinault people, *The Quinault Indians,* Ronald Olson was clearly disinterested in plant-use traditions, seldom inquiring about the topic with the tribal interviewees of his day and devoting only a few pages to the topic. In-house publications by the Quinault Indian Nation offered important corrections on many points, but did not focus specifically on the topic of plant knowledge.

In response, Deur and James collaboratively developed a long-term study of past and present Quinault plant knowledge. They have carried out interviews with knowledgeable tribal members about their traditional uses, views, and values relating to plants. Participating tribal members were generous with their vast experience with the traditions, making contributions that are mentioned throughout the book. Additionally, this research has involved a careful review of what was written about Quinault plant-use traditions in books, articles, and unpublished accounts. Together, these older sources were synthesized with the contributions of recent interviewees, providing a very detailed picture of Quinault ethnobotany as it endures today.

In the pages that follow, we do not identify every species mentioned by elders and written accounts. Such a book would be huge, and too bulky to serve as a field guide. Instead, we focus on the "cultural keystone" species—plants that have been especially important in the cultural traditions of Quinault and other nearby tribes, past and present.

In this book, these plants are organized into three chapters: "Trees of the Forest and Forested Riparian Areas," "Shrubs for Food and Medicine," and "Other Important Plants of Meadows, Wetlands, and the Forest Floor." Each chapter begins with the most prominent species of each category, and each offers enough detail to help even novice plant users get oriented and begin harvesting and using native plants. Plant entries include information on traditional uses of each species, as well as informational sections for each plant: "What It Looks Like," "Where to Find It," "When to Gather," and "Traditional Management." Some entries, especially for plants that can be poisonous or harmful, also include a discussion titled "Cautions." Each chapter is illustrated with photographs and other images. Anyone wanting more information on these plants can consult online and printed plant guides.

The contents of this book are also a gift from past generations of knowledge-holders to knowledge-holders of the future. Quinault people have long dwelt in an abundant land, its natural resources sustained for countless generations by the careful, patterned, sustainable use of their ancestors. Though new challenges threaten to undermine Native American plant use and the integrity of the plant communities on which the ancestors depended, hope remains. By respecting this natural and cultural inheritance, and by continuing to learn of it and embrace it, countless future generations of Quinault people will be sustained, along with their new neighbors in the Pacific Northwest—this place we together call home.

A Note about Using Plants

Do not collect, consume, or use any of the wild plants identified in this book without first consulting a health-care professional. The information provided in this book is informational only and is not intended to diagnose, prescribe, prevent, or treat any illness, injury, or health condition. Always consult a health-care professional when suffering from any health condition, disease, illness, or injury, or before collection, use, or consumption of any traditional health remedies. Any person who collects, consumes, or uses plant materials does so at his/her own risk.

The many individuals and organizations who have contributed to this book all have exercised great caution in providing an accurate reflection of plant use as practiced by Quinault people within the pages that follow. The food and medicinal uses described in this book are reported to be longstanding practices of the Quinault and other tribes of the Pacific Northwest coast. Still, there are dangers to wild plant use for food, medicine, or other purposes. Some plants are not safe for some individuals, particularly those with certain medical conditions, those taking certain medications or supplements, or pregnant women. The danger of certain species may vary from plant to plant depending on location, timing, the genetics of a particular plant community, and other factors. As with any natural product, plants can be toxic if misused. The chemicals in some species can counteract important medications. The safety of wild plant consumption continues to be a subject of scientific study and debate, and information reported at one time on plant safety may subsequently be superseded by new scientific studies and findings. There are also risks of mis-identifying plant species. We identify many known risks of plant consumption in the pages that follow, but those warnings are only a starting point. Readers are urged to carefully cross-check the identities of plants with other guidebooks and the abundance of online resources to aid in plant identification.

We discourage collecting, consuming or otherwise using the plants as described in this book unless you have verified with your healthcare professional that they are safe for you. If you receive approval from your health-care professional, you are responsible for following best practices in the collection, consumption, or use of plants, including but not limited to clearly identifying plants, keeping plants away from children and animals, and storing plants in a safe location.

The author, the Quinault Indian Nation, Oregon State University Press, their officers, agents, trustees, and members make no guarantees as to the safety of collection, use, or consumption of any plants, and disclaim all liability and responsibility for any consequences resulting from the use or reliance upon the information contained in this book, and for any health problems, consequences or symptoms which may arise from contact with, use of, or the consumption of any plant described in this book. If you have any doubts whatsoever about the safety or identity of a particular plant, we strongly urge you to avoid it.

Gifted Earth

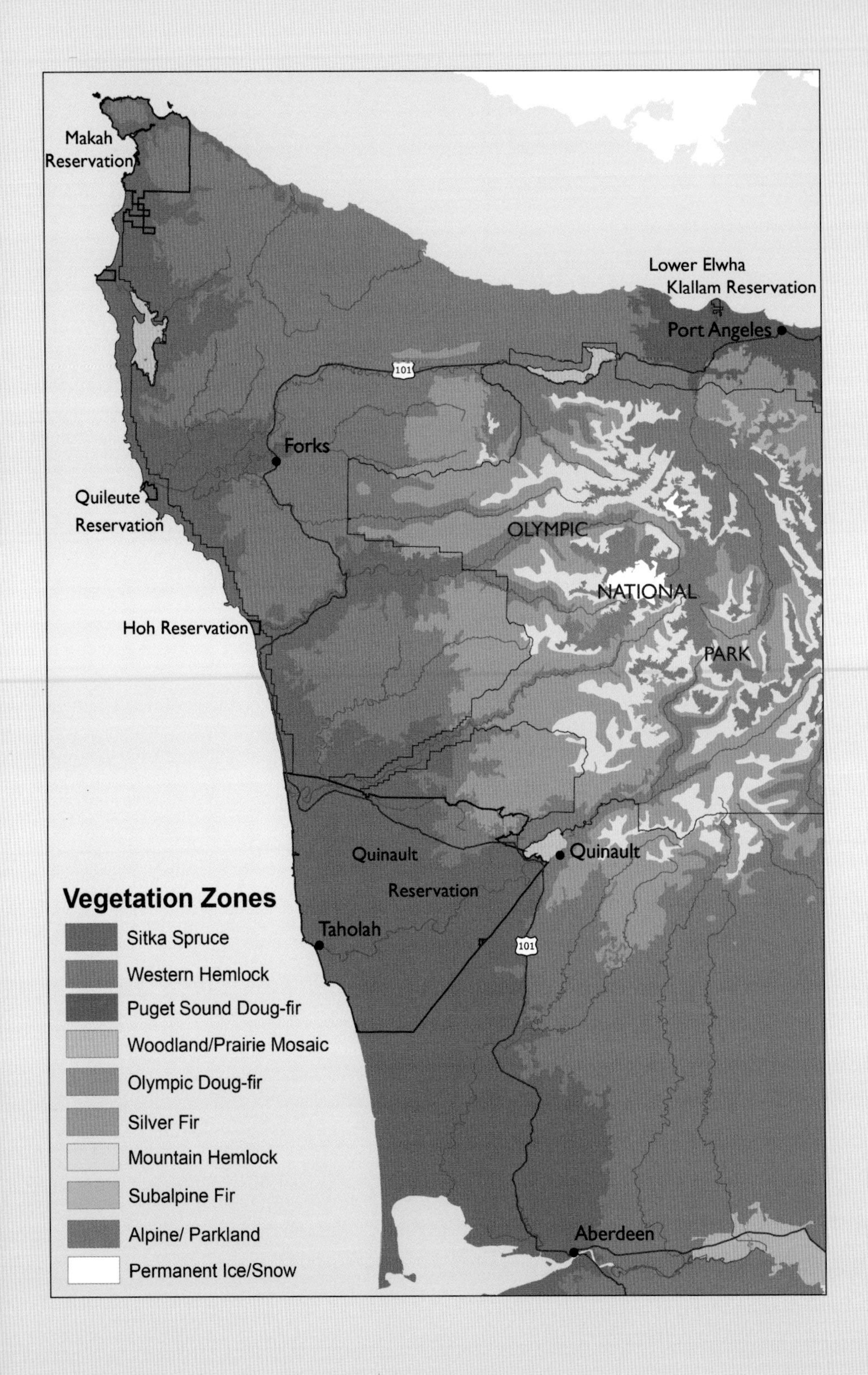
Makah Reservation
Lower Elwha Klallam Reservation
Port Angeles
101
Forks
Quileute Reservation
OLYMPIC
NATIONAL
PARK
Hoh Reservation
Quinault Reservation
Quinault
Taholah
101
Aberdeen
Vegetation Zones
Sitka Spruce
Western Hemlock
Puget Sound Doug-fir
Woodland/Prairie Mosaic
Olympic Doug-fir
Silver Fir
Mountain Hemlock
Subalpine Fir
Alpine/ Parkland
Permanent Ice/Snow

Ethnobotany in the Land of the Quinault

On the wet outer coast of Northwestern North America, where the Olympic Peninsula meets the Pacific Ocean, live the Quinault. They have lived along this stretch since time immemorial, their homelands spanning from jagged rocky mountaintops, through dense and towering rainforest, down dancing rivers to salty bays, ocean beach, and open sea. On the western flank of the Olympic Peninsula, in and around lands now within Olympic National Park, the territories of the Quinault and neighboring tribes hold biological riches famous the world over.

Through millennia living amid such abundance, the Quinault and their neighbors developed intimate connections to living things in this landscape—a relationship that has been carried from generation to generation. Plant life was fundamental—a source of foods, medicines, materials, and spiritual sustenance. The ancestors experimented with plants, had plant uses revealed to them, and exchanged this knowledge with neighboring tribes since long before European peoples dreamed of this coast. The vibrancy of the land, together with knowledge of how to use its great inheritance, were gifts bestowed by the Creator and transmitted by ancestors to each new generation. These gifts protected and sustained the people, that they might survive and endure into present day.

In spite of myriad changes over the last two centuries, plant uses and values persist and are recalled by knowledge-holders within the Quinault Indian Nation and neighboring tribes of the Washington's Pacific coast. In the pages that follow, these knowledge-holders share some of their accumulated knowledge with the wider world.

Opposite: Major vegetation zones of the western Olympic Peninsula, homeland to the Quinault and neighboring tribes.

Shared openly and enthusiastically, this information is a testament to the wisdom of the ancestors and proof of the tribe's cultural endurance. Yet, there are more urgent reasons to share. Elders perceive that the fates of residents, Native and non-Native, on this land are complexly intertwined and increasingly interdependent. Thus, the Quinault also share the plant knowledge in this volume in hopes it might inspire non-Native peoples living in the Northwest—those who seek balanced and sustainable modes of living—so that ancient environmental and cultural gifts might be respected widely by the region's human inhabitants, so that these things might thrive, sustaining future generations.

Thus, this book is a gift to the outside world. It is a gift that, by Quinault tradition, must be reciprocated with equal or greater gifts in return. How does a reader repay the full value of this book? This debt is repaid through respectful treatment of Western Washington Native peoples, their lands, their waters, all the culturally important plants, and all living things native to this place. Reading and using this book is not a passive exercise; those who read it are expected to honor this bargain. If they learn from this book, apply its lessons, and show due respect, all might hold greater hopes for long-term success and joint survival in this place that we call home.

Enduring Relationships, from the Summits to the Sea

The ancestral Quinault homeland includes the western slopes of the Olympic Peninsula—especially the riverbanks, bays, and ocean shores. In this book, however, we speak of not only the Quinault people, but of the Quinault and their neighboring tribes, representing Native American peoples and homelands along much of Washington State's Pacific Ocean coast. This reflects in part the fact that the Quinault Indian Nation is a remarkably diverse tribal community, embodying the traditional knowledge of tribes along the entire ocean coast of the Pacific Northwest. Their membership consists of descendants of many tribes from the northwestern Olympic Peninsula to the northern Oregon coast who were relocated to Quinault in the nineteenth and early twentieth centuries—including many Chinook, Chehalis, Quileute,

Queets, Hoh, Tillamook, Clatsop, and others tribal descendants, in addition to Quinault proper.

The modern Quinault people, with all of these roots, continue to live on the ocean coast. Their members are among the foremost knowledge-holders on ethnobotanical matters for the Pacific coastal tribes hailing from far Western Washington and even northwestern Oregon. Individuals descended from each of these tribal communities have contributed to the current volume, giving this book breadth beyond the knowledge of a single tribe and enhancing its value as a guide to traditional plant knowledge along the entire Pacific Northwest ocean coast.

Reflecting this expansive heritage, the traditional homelands of modern Quinault people at minimum span almost the whole of far-western Washington, from the northwestern Olympic Peninsula to the mouth of the Columbia River. They front more than two hundred miles of Pacific Ocean coastline, and stretch from the 9,726-foot summit of Mount Olympus to the sea. Together, these territories contain an astonishing diversity of natural habitats that have long sustained ancestors of the Quinault's constituent tribes. These homelands encompass some

THE QUINAULT SEASONS

April	p̓angwahʔám huhnshaháʔ—Time when the geese go by
May	p̓anjulashx̣uhtltu—Time when the blueback return
June	p̓ankwuhlá—Time of the salmonberries
July	p̓anklaswhas—Time to gather native blackberries
August	p̓anmuuʔlak—Time of warmth
September	t̓s okwanpitskitlʔak—Leaves are getting red on the vine maple
October	p̓anʔsilpáulos—Time of autumn
November	p̓anitpuhtuhkstista—Time when the clouds are covering
December	p̓anklich—Time of darkness
January	p̓anpamás—Time of cold
February	p̓anlaleah-kilech—Time of the beach willow (*Salix hookeriana*)
March	p̓anjáns—Time of the berry sprouts

Cedar dugout canoes and cedar clothing are among the many plant materials that have sustained Quinault life along the rugged, rain-soaked outer coast of Western Washington.

of the largest and richest estuaries on the west coast of North America; famously lush temperate rainforests containing some of the world's largest trees; small coastal prairies that tribal ancestors traditionally kept cleared using fire; a variety of lakes and freshwater wetlands; comparatively dry interior forests and grasslands; subalpine forests and meadows; rocky mountain slopes and peaks. Each of these habitats contains distinctive plants and plant communities. So too, each of the habitats holds unique importance, providing specialized plant harvest opportunities and venues to the ancestral Quinault. Each traditionally utilized plant came from its own particular habitat and place, from the Olympic Mountain summits to the sea. Each year, the ancestors visited these places in seasonal cycles to harvest the plants needed to sustain each aspect of life.

The coastline offered its own unique possibilities. From the rocky shorelines and the base of headlands, ancestors gathered edible seaweeds at low tide, while edible eelgrass rhizomes could be found in the bays. The salt-tolerant marshes lining bays and tidally influenced rivers were especially productive. Here, the ancestors harvested edible intertidal roots such as springbank clover and Pacific silverweed, while gathering sweetgrass, sedge roots, cattail, and other plants for weaving baskets, mats, and other items.

The deep coastal forests provided an abundance of trees, ferns, berries, mosses, and many other plants well-suited to the cool, damp shade and the moderate coastal climate. It was within these forests that the ancestral Quinault found their cedar wood and bark, though many other tree products were gathered as well: for example, the medicinal sap of Sitka spruce, the hard wood of yew for tools, or the springy vine maple limbs for making baskets. Berries such as salal, red huckleberry, and evergreen huckleberry abounded in the understory, as did ferns used as foods, medicines, weaving materials, and for other purposes.

On the shady, sun-dappled forest floor there are edible greens too. Tribal members recall gathering wood sorrel (*Oxalis oregana*), for example, with its shamrock-like triad of leaves. Though toxic in large quantities, these plants make a refreshing snack and tribal members gather them to eat or just to chew, savoring the sour flavor, often compared to apple or lemon. Some tribal members called these plants "bear candy," as bears are known to enjoy and be drawn to patches of wood sorrel in the deep forests of Quinault country.

The clearings within these forests were especially rich places, kept open by the repeated use of fire. On these "prairies," the Quinault harvested plants like Indian tea, the sweet edible bulbs of camas, tasty wild cranberries, and bracken fern with its edible roots and "fiddleheads." On the margins of these clearings, along the rivers, and in many other open places was an abundance of trailing blackberry, evergreen huckleberry, and many other edible species. In other burned settings, such as burned forest understory, Quinault people maintained productive patches of blackcaps and other tasty berries. These places were tended like gardens, revisited year after year and cultivated with great care.

In certain clearings the Quinault traditionally harvested vast patches of beargrass—one of the most popular basketry plants used by Northwest tribes. Though ordinarily a plant of the high mountains, beargrass was pushed by glaciers to these low-lying areas long ago. Here,

Even fishing and hunting gear, and items such as the modern hand-woven clam baskets pictured here, are made of plant materials.

Native people maintained beargrass patches through regular use of fire, enhancing its output considerably. A source of Quinault wealth, the ancestors piled beargrass high in canoes and paddled to intertribal gatherings on the Columbia River, where it was a popular trade good.

In autumn, ancestral Quinault traveled high into the Olympic Mountains, establishing camps in places like Enchanted Valley where women, children, and the elderly harvested oval-leaf blueberries and mountain huckleberries as men fanned out to hunt in areas nearby. People ascended higher still—far into subalpine areas and up to the mountain peaks, for plants used for medicinal and spiritual purposes that could be found at extreme elevations. Recent research suggests that such plants do indeed have unique chemical properties tied to the climate and soils of high elevations, in addition to being linked to places of profound cultural and spiritual importance.

While the cumulative botanical inheritance of these coastal homelands is tremendous, so too is the depth of plant knowledge carried forward by the Quinault and neighboring tribes. Together, and in almost every imaginable sense, a rich botanical and ethnobotanical inheritance sustained the ancestors of the Quinault. Tribal oral tradition and archaeological evidence agree: plants have always been important in the lives of people within

Quinault and neighboring tribes. Based on the teachings of elders and a review of available written materials, we know that the Quinault proper regularly utilized over one hundred species of plants, and used at least two to three times that number if we account for specialized uses and occasions. At one time, almost every aspect of life involved the use of plants. Roots, berries, shoots, greens, and other plant products were staple foods, integral to healthy diets that also included an abundance of fish, shellfish, and meat. Plants were used in medicines to cure many different ailments, and most spiritual and ceremonial traditions involved plant materials in various ways.

Plants also were used in the manufacture of almost every object included in the Quinault household: houses, clothes, tools, baskets, canoes, and innumerable other goods. Even hunting and fishing has traditionally depended on plants—with spears, bows, arrows, weirs, and other gear made largely of wood and other plant materials.

Each wood has its own qualities, and traditional Quinault craftspeople made the most of these distinctions, traveling to special places to find woods with exactly the properties desired. From the wood of Western redcedar and other trees, the ancestors built houses, canoes, utensils, tools, and most other durable items used in the Quinault household. The fibrous inner bark of the cedar tree, peeled and pounded cottony-soft, is traditionally woven into clothing, hats, mats, and regalia.

Quinault and their near neighbors have long been famous for a rich weaving tradition—shaping baskets, mats, and other woven items not only from cedar bark but also from native beargrass, sweetgrass, sedges, ferns, woods, and other plants. A long and lively trade in plant materials and plant products, such as baskets, linked the Quinault to other tribes; and unique access to materials like beargrass contributed to Quinault tribal wealth.

Today, the Quinault and neighboring tribes continue to use native plants in myriad ways. Berries, berry sprouts, Indian tea, and many other plant foods still sustain the people, as healthy and symbolically potent components of the larger diet. Plants also continue to be used as medicines, and medicinal knowledge of plants expands and evolves as tribal members address once-rare medical conditions ranging from diabetes to cancer. Many families continue to gather plant materials such as cedar bark, cattail leaves, and sweetgrass, producing beautiful baskets and other woven goods. A rich carving tradition also persists and evolves over time, with new cedar canoes, carved poles, masks, rattles, and other items as centerpieces of modern Quinault ceremony and expressive culture. Together, these practices are part of the living ethnobotanical tradition of the Quinault people—the enduring knowledge, use, and management of plants by a people who share a deep cultural tradition and ancient ties to their Western Washington homelands.

"[Quinault people traditionally burn prairies] towards the end of spring. Where the prairies were just getting kind of dried out a little bit. And where it would just go up, buffer up against the trees. And it would just stop right there. And all the vegetation inside will all sprout up, oh, in a matter of weeks you see a bunch of green sprouts popping out of the ground. . . . There's a lot of berries that come up [and] camas . . . Comes back better. All the nutrients just fall to the ground and everything sprouts up, all the seeds that were there. Some of them need to be burned in order for them to sprout . . . and the Indian tea . . . Guess it would make a better tea . . . More nutrients in it."—Dennis Martin

The Wisdom and Logic of Traditional Plant Management

Plants abound in the homelands of Quinault people. Yet this natural abundance and biological diversity reflects, in some part, ancient traditions of stewardship among the Quinault and other Washington coast tribes. The ancestral Quinault were active managers of the landscape, employing many strategies to ensure the long-term health and productivity of culturally important plant and animal species, and showing respect to the Creator, to creation, to culturally important habitats and places, and to the individual plants harvested. Even long ago, the Quinault had "anthropogenic" plant communities—those that were significantly the product of human management, even if the individual plants and habitats were native in origin.

To enhance plants and their habitats, the ancestors did many things. They harvested plant parts, such as cedar bark, selectively, so as not to kill the plant. Only one or two strips could be pulled from a cedar tree before it was allowed to recover fully and grow new bark over the scar—a process that could take centuries. Likewise, they harvested plant materials such as camas bulbs selectively, to avoid overharvesting or the elimination of whole patches—a genuine threat on the outer ocean coast where camas was patchy in distribution and relatively rare. Reckless harvests would be felt almost immediately by individuals and communities. Thus, the incentives for conservation were clearly material, as well as cultural and cosmological.

More active interventions were also part of traditional Quinault habitat management. Past and present Quinault basketmakers have used pruning to increase plant output or, for example, to produce long, straight shoots of willow and vine maple so prized in weaving. Pruning these plants low on the stalk produces long, straight shoots in the year ahead. At times the ancestral Quinault appear to have even seeded plants, transplanted plants, or weeded out competing species to foster culturally significant species—with camas and other relatively rare but important species as a focus of many such practices. Fertilizing seems to have been used as well, as when fish waste was spread around nettles and other culturally important species.

"I've done it, I've gone out and burned, burned the blackberries. Because once the blackberries get burned, it goes down into the roots, telling them, 'Hey, you got to get some seed out there because we're going to be gone!' So the next year, there's going to be the whole patch, just packed!"
—Harvest Moon

Yet among the many techniques used to enhance plant output, none left such broad or enduring effects as the use of fire. In myriad settings, the ancestors burned vegetation, creating openings in the dense forest filled with culturally important species, and even burned the understory of forests to aid in the output of culturally preferred plants. Recent research confirms that fires were essential, removing competing vegetation, fostering seed germination, and releasing key nutrients into the soil. The scale and methods of burning varied, reflecting the habitat and the target species involved. Fires on wetland prairies maintained open, grassy patches full of Indian tea, camas, bracken fern, wild cranberry, and other plants; these places also served as predictably good hunting grounds. Lands within prairies were loosely owned by families, creating incentives to tend the plant habitats in their control. Nearby, Quinault people burned beargrass to enhance productivity and keep competing vegetation at bay. By burning forest clearings, and even the forest understory, ancestral Quinault enhanced the output of edible berries such as huckleberries and trailing blackberries and blackcaps. And, in the Olympic Mountains, Quinault harvesters also maintained and expanded mountain huckleberry patches through the regular burning of forest edges and understory.

Through all these mechanisms, from selective harvesting to burning to hands-on plant cultivation, the ancestral Quinault left a significant imprint on a landscape that was erroneously perceived as "wilderness" by arriving Europeans. The effects on the landscape and its species were detectable to knowledgeable or careful observers—not only as a reflection of sophisticated and sustainable Quinault food production methods, but also as a manifestation of the deeper cultural and spiritual values of Quinault people.

"They were an especially gifted people. There was something in this earth that was put there by someone; they honored everything. I talked to the real old ones who were around eighty to one hundred years old at the time I was a young person. . . . They figured that was quite an honored privilege, put on for your livelihood. They respected that power; my people respected God's power to create. They wouldn't go to the extreme of being greater than God, to put things on this earth. They said if you try to do that, it's going to disappear from the earth and you aren't going to have it. That's what they put into my heart and my mind and that's the way I grew up and the way my peers taught me. They didn't waste anything that was of high value for survival."—Horton Capoeman

Each plant cultivation method can be understood only within the context of these deeper cultural values. These values are evident in the potlatch—the ceremony in which the Quinault and all other tribes of the Northwest systematically honor people, gift them, and repay debts. By doing this correctly, one can maintain necessary balances within and between human communities; by doing this correctly, one can prevent the economic, social, psychic, and spiritual "imbalances" caused within and between communities by unpaid debts, poverty, theft, and unchecked human greed. Likewise, traditional plant cultivation methods reflect Native American communities' appreciation of long-term mutual indebtedness between "the community of plants" and "the community of plant users." The ancestors of these plants cared for the ancestral Quinault and, since time immemorial, harvesters have sought to honor this fact when gathering plants, showing appropriate respects. Traditionally, harvesters have conducted blessing ceremonies and sometimes left offerings, as they show thanks for the life and vitality sacrificed by harvested plants for human well-being. If traditional management methods such as burning and selective harvesting cause plants to come back more abundantly, harvesters understand that the plants, in their way, appreciate the intervention. If they come back in this way, harvesters recognize that they have successfully repaid debts to the plants for the bounty of food, medicines, materials, and other gifts they continue to provide to Quinault people. If they take care of them in this way, the plants will continue to take care of their human harvesters.

Traditional burning clears away brush, releases nutrients to the soil, and enhances the growth of many culturally important plants within small "prairies" still found today along the ocean coast rainforest.

Even today, traditional management involves showing respect to important plants by helping them thrive, harvesting selectively, and killing only when necessary. For example, Quinault gatherers traditionally take only a few Indian tea leaves from each plant so that no one plant is critically harmed. Similarly, traditional harvesters only take three interior leaves from any beargrass plant—never so much that gathering might kill plants or eliminate whole patches. On the tide flats, Quinault harvesters pick sweetgrass carefully, so as to neither kill the plant nor damage harvested stalks. This shows respect to the plants and, by ensuring a sustainable supply, shows respect for future harvesters. The same values come into play in the traditional use of tree products. In harvesting materials from cedar trees, the Quinault historically seek to take only what is needed—peeling strips of cedar bark or even removing boards from the sides of ancient cedars while the tree is left standing and alive. Today, these "culturally modified trees" are sometimes seen in Quinault country forests with scars on their sides—some relatively new, some dating from long before Euro-Americans arrived on these shores. These are living artifacts of deep, respectful philosophical traditions and of the mutually supportive relationships between Native peoples and the plant communities of their homelands.

Other anthropogenic plant communities can be seen in the Quinault homelands to the present day. Many productive berry-picking areas are the product of repeated human burning and other forms of traditional management. Even where the landscape has not been actively maintained for many years, one still sees clearings, unexpected concentrations of culturally important plants such as camas and berries, and other signatures of this enduring relationship between Native peoples and the plant communities of their homelands. Like culturally modified trees, the prairies and berry patches remain in the lands of Quinault people—artifacts of traditional management techniques and vibrant philosophies.

Tribal members observe many of these practices today, along with the traditional philosophies supporting them. Increasingly, the Quinault Indian Nation and other tribes explore ways to incorporate traditional methods and philosophies into modern resource management on their reservations and beyond.

The Displacement and Endurance of Quinault Plant Traditions

Plant use changed in the wake of Euro-American settlement and the forced relocation of Native American people throughout the region. The Quinault Treaty, too, changed many things. Signed in 1855–1856, this treaty relocated tribes of Washington's Pacific Ocean coast, except Makah, to the Quinault Indian Reservation, located within the traditional territory of the Quinault proper. Some families moved to Quinault from other places and other coastal tribes, such as Chehalis, Chinook, Hoh, Queets, Cowlitz, and Quileute, contributing to the richly multi-tribal character of the Quinault Indian Reservation. Some resisted relocation and stayed on their traditional lands, while others moved to Quinault only for a time before returning to their original homes. A few tribes from northwestern Oregon such as Clatsop and Tillamook—their treaties with the United States being unratified by Congress—also ultimately found their way to Quinault. Often this relocation was shaped by kinship and other connections

"People . . . use a lot of devil's club. . . . They use it for arthritis. They mix it into a salve and put it on arthritis. And for the skin. They can find more uses for it now than they did back when they used it [in old times]."—Dennis Martin

The Quinault Indian Nation Roger Saux Health Center has many healthy recipes for the use of wild plants, many of them updated for modern tastes and technologies. One of many examples is . . .

Rosie's Cold and Flu Tea

Gather equal measurements of the following dried herbs:

½ cup wild mint leaf
½ cup elderberry flowers
½ cup yarrow leaf & and flower
3 teaspoons dried orange peel
3 teaspoons dried lemon peel
3 teaspoons dried lime peel

Mix all ingredients thoroughly. Store in a sealed jar and use as needed for a soothing cup of tea.

Add a touch of fresh lemon juice and honey to taste.

Courtesy Rosie Carpenter-Reed
Quinault Diabetes and Chronic Disease Prevention Program

with the Chinook. Each community that converged on the Quinault Indian Reservation brought with them interrelated but subtly distinctive plant knowledge and plant-gathering traditions.

Plants were clearly of concern to the leaders of the multiple tribes who took part in the 1850s negotiation of the Quinault Treaty. Though US negotiators rebuffed many of their proposals for enduring land and resource rights, these leaders still argued successfully for continued plant use and access beyond the Quinault Indian Reservation. Thus, the Quinault Treaty not only reserved to the Quinault people fish harvesting rights at all "usual and accustomed grounds and stations," but Article 3 of the treaty also reserved "the privilege of hunting [and] gathering roots and berries" on "all open and unclaimed lands." Exercising this privilege, the Quinault proper still continued to access the full range of traditional gathering areas. So too, each of the peoples who converged at the Quinault Indian Reservation from other tribal lands retained certain ties to their original homelands and their ancestral plant communities, even generations after their displacement to Quinault.

"Indian tea. . . . Supposed to heal everything! I still drink it when I need something [for] relaxing or something."
—Clifford "Soup" Corwin

As non-Indians began to occupy the lands beyond reservation borders, tribal members continued to visit their traditional gathering areas—especially the undeveloped

Quinault people continue to use many traditional plant foods regularly, while reacquainting themselves with common plants, such as springbank clover (*Trifolium wormskioldii*), that were important to the diet and culture of the ancestors.

spaces remaining within their increasingly developed Western Washington homelands. As non-Indian people moved into the Aberdeen-Hoquiam area and developed historical sweetgrass harvest areas along the intertidal zone, for example, harvesters adapted, using some of the old gathering sites while moving to second-tier harvest areas elsewhere along the shores of Grays Harbor. As non-Indian agriculturalists occupied off-reservation prairies, tribal use of the prairies remaining on the reservation became even more important. In gathering areas throughout Quinault country, tribal plant harvesters contended with the increasing effects of expanded non-Native settlement. They were forced to navigate reduced opportunities associated with poaching, pollution, limits on access or harvest, restrictions on traditional land management practices, and other modern obstacles. In each case, families have adapted. Simultaneously, the growing ease of transportation within and beyond the Quinault region has allowed for access to plant materials from other tribal territories within the Pacific Northwest and North America beyond—allowing tribal members to integrate novel native plant materials into very ancient basketry, carving, medicinal, and food plant traditions.

Today, members of the Quinault Indian Nation continue to harvest and use plants for many different reasons. Both on- and off-reservation, plants are gathered for use as foods, medicines, and materials. For some, the use of native plants is an important part of tribal identity, and a big part of what it means to "be Indian" in modern times. The berries consumed today are the direct descendants of those that fed their ancestors; therefore, consuming even a few of these berries can be a potent way of reconnecting with the traditions and homeland of their forebears. Tribal art and ceremony still involve many plant products as well, especially the wood and bark of redcedar, which remains a cornerstone of Quinault cultural practice. At Quinault and in neighboring tribes remain skilled specialized canoe builders and pole carvers. Some gatherers also peel bark for hats, regalia, baskets, and many other cultural uses. For reasons cultural, social, spiritual, and even economic, many families still gather materials for the production of baskets. Skilled Quinault basketmakers harvest

Quinault families still gather camas bulbs and other plant foods, as well as plant medicines and craft materials, from prairies long managed by their ancestors.

sweetgrass, beargrass, cattail, vine maple, willow, cedar bark, maidenhair fern, spruce root, and other species, producing woven items for sale and trade, for personal use, or as gifts for esteemed family and friends.

Special gathering places also endure, especially on the reservation. Many families maintain their own favorite and secret picking spots. The places visited by Quinault tribal members for plant gathering today are "cultural sites" even in the absence of physical traces on the landscape. These places, and the plants harvested from them, are important to the survival of the living culture. The survival of Quinault canoe-building traditions depends on a steady supply of large cedar trees. The survival of Quinault basketry traditions depends on a steady supply of materials like sweetgrass and beargrass. Sometimes, tribal members' ability to withstand illnesses or other trauma depends significantly on their ability to use traditional foods and medicines. Recognizing this, the staff of the Quinault Indian Nation works to help protect the integrity of these cultural resources for the benefit of all

Quinault—those living today and those yet to be born.

Evidence of the ancestors' special rclationships with plants abounds in Quinault country. Though many prairies have disappeared with the loss of traditional fire management and the rise of competing land uses, a number of prairies remain. In these places, families have gathered together, camped, harvested resources, and socialized for generations. Not only do the clearings contain signature anthropogenic vegetation, but the soil contains lenses of old charcoal and other evidence of long-term use and management by the ancestors. These places offer powerful connections to ancestors who sustained the habitats and were sustained by them.

Deep in the forest, in many parts of the Pacific Northwest, one can still find culturally modified cedar trees, bearing scars where strips of bark or planks were removed long ago. They are living artifacts of ancestors' cultural activities and values. Such trees are also powerful reminders to modern people—reminding observers not only of Native American plant gathering at that exact spot, but of the ethical and ecological logic behind traditional selective harvests. For many tribal members, beholding

Devil's club (*Oplopanax horridus*), among the most potent and valued medicinal plants of the coast, raises its sprawling prickly leaves toward the dappled sun of the forest understory.

Quinault traditional plant specialist, Harvest Moon, gathering sweetgrass (*Schoenoplectus pungens*) for baskets.

a culturally modified tree is a profound experience. Such landmarks—verifiably touched by the ancestors of a very different time—are considered powerful, even sacred, by many tribal members.

For modern tribal families, access to native plants also provides immediate and tangible benefits. The food use of plants remains widespread, with berries being especially popular. Tribal members gather and eat all species of huckleberries and blueberries available on Washington's Pacific Ocean coast, as well as salmonberries, salal berries, blackcaps, blackberries, thimbleberries, elderberries, wild cranberries, wild strawberries, crabapples, and others. Some still eat edible bulbs and roots, such as the sweet bulb of the camas lily, though in small quantities because of its limited availability. Tribal members also eat certain edible leaves and shoots, such as the soft springtime shoots of salmonberry and thimbleberry. Harvested in wetlands and prairies, Indian tea is still a popular and healthy beverage. Families gather these plants as part of special social events, and the Quinault Indian Nation and neighboring tribes host tribal events and share recipes meant to foster continued uses of traditional plant foods.

Regardless of their goals, tribal members appreciate that, when used in traditional ways, these plants often provide healthy and inexpensive alternatives to introduced foods or manufactured goods that must be purchased and shipped from elsewhere. Traditional plant foods are generally high in vitamins and minerals, antioxidants, fiber, and many other nutrients that sustain good health. Combined with salmon and other traditional fish and game, they contribute to a remarkably healthy diet. To many modern Quinault, every native plant used in lieu of an introduced food or medicine is seen as a minor victory against the negative effects of cultural change, economic challenges, and the erosion of the traditional diet. Trading out even a small amount of processed food for traditional plant foods is understood to yield potential health benefits; in this way, traditional foods are healing, and serve as both "food" and "medicine."

Medicinal plant traditions persist as well. The list of plants used for medicine is very long: popular plants include cascara, devil's club, coltsfoot, fireweed, salal leaves, wild mint, various ferns, and even introduced species like dandelion. Some tribal members harvest spruce pitch for medicinal use, but also for use as a sealant in traditional wooden crafts. Medicinal plant uses continue to develop and evolve as tribal members discover ways of reducing the effects of increasingly prevalent ailments like cancer, diabetes, antibiotic-resistant infections, and cardiovascular disease. These uses are sometimes learned from friends and family in other tribes, or are revealed to tribal members through experimental use and recombination of time-honored ingredients, spiritual practice, and dreams. The use of plants for medicine is probably the most rapidly changing aspect of traditional plant use, reflecting this growing body of knowledge and a practical response to the new threats of the twenty-first century. The rapid evolution of medicinal plant use is a sign of the enduring vitality of Quinault plant use traditions, and its likely relevance well into the future.

Trees of the Forest and Forested Riparian Areas

Crossing the Quinault River at Taholah, circa 1912.

Thuja Gigantea.

Gigantic Arbor Vitae. *Thuia gigantesque.*

Western Redcedar

čiitəm | *Thuja plicata*

Traditional Uses

Among Northwestern coastal tribes, western redcedar is the most important plant by any measure. No tree has been as thoroughly woven into social, economic, cultural, and ceremonial life. At every step along life's journey, the cedar accompanied ancestral Quinault people, sustaining them from birth, when they were wrapped in cedar bark baby clothes and secured in a cedar bark cradle. It sustained them through lives spent in cedar plank houses and when traveling in cedarwood canoes. Finally, it sustained them at death, when their bodies were wrapped in cedar bark blankets in cedar coffins. Even today, the tree maintains profound importance in both mundane and sacred domains.

Cedar Wood

Cedar wood is famously fragrant, fine-grained, and easy to carve. Beyond this, it is rot-resistant—so much so that logs can lie on the forest floor decomposing almost as long as they previously stood aloft and alive. The Quinault historically crafted most daily necessities from cedar. And traditionally, canoes are carved from cedar logs with tremendous care and artistry. When seeking a canoe log, carvers look for trees with straight, tight grain and few limbs, knocking on the trunk and sometimes carving a small hole to determine the solidity of the wood. Moreover, trees producing more cones and seeds are deemed best for canoes. Historically carvers sometimes sought out canoe trees near rivers and bays so that, once felled, the logs could be partially worked on-site then floated to the village for completion. Yet, like other tribes

Strips of its bark removed for cultural uses, a cedar tree remains upright and alive.

"Some people go along the river. And they were long and straight, the [cedar] root. They had to be just so. Some people looked around in the woods and they used to look for the really old trees that laid there, you know they're red now, turning to dirt, I guess. You'd find the roots there on the log—an old, old log. You'd pull that out and they liked it because they were long. . . . I imagine there might be some on the Quinault River. Nice, long, straight ones. Not big ones, but bigger than my fingers. . . . I'd take them home and clean them. Then you'd split them. And then you'd fix them the way you want the root—dry it straight or folded or something. . . .They have to be long and straight. The water keeps them that way."
—Katherine Barr

of the region, the Quinault especially valued trees from higher elevations for their tight grain, sometimes seeking such trees despite the acute challenges of transporting downed logs to sea level. Windfall and drift logs were also used when available.

Even long ago, cedar tree fellers were specialists, recruited and compensated for their skill. With demonstrations of deep ceremonial respect, they built a platform around the trunk, chipping away with adzes and sometimes fire. Once the tree was felled, it was commonly floated to see if one could discern its natural upright position. Carvers then worked a rough canoe shape from the log, using elk bone chisels, adzes, wedges, and sometimes fire. When a cedar log was reduced to a canoe "blank," it was floated to the village for finish work. In some places on the Olympic Peninsula, unfinished canoes have been reported still hidden under the moss of the forest floor, far from any village.

Fire was used to deepen the interiors of the canoes. When the canoe was approaching completion, carvers commonly filled it with water and fire-heated stones; this made the wood flexible, allowing carvers to mold it precisely to a desired shape and to complete fine-grained carvings, finalizing the elegant finished canoe form, each canoe customized for its intended use. To the painted and sealed canoe hull were added thwarts and seats, fastened with cedar roots and "withes" made of flexible branches that could be used to make canoe repairs. Many items—such as cradles, coffins, and feast dishes—were sometimes crafted like small canoes, using the same dugout techniques on a smaller scale.

House planks were made from western redcedar as well. Planks could be split from whole cedar logs, but were also traditionally split with wedges from living trees—reflecting both the difficulty of tree-felling and a desire to not kill a tree unnecessarily. Charred and adzed to the ideal thickness, these planks were historically used to build the walls and roofs of most houses around a cedar pole frame, often with cedar plank floors, cedar doors, and cedar benches and platforms lining the house's interior walls. In times of conflict, villages were historically encircled by fences or palisades made in whole or part from cedar posts. Inside traditional homes were many cedar boxes made by grooving and then bending a single plank at right angles in three places to create the four sides of the box, then attaching the ends with pegs of spruce or cedar where they met the base of the box. Cedar boxes have been used to store almost everything from food to clothing to ceremonial regalia.

Beyond these fundamental uses, cedar wood has been used to manufacture many other items: paddles and punt poles, bows and arrows, and digging sticks, fish traps, and duck decoys. Cedar buoys and floats carved to look like ducks were sometimes tied to harpoons or fishing gear. Cedar has also been used to carve "totem" poles, rattles, and all manner of ceremonial items. Spindle whorls and weaving tools, designed to make cedar bark mats and clothing, were also made of cedar wood in addition to other tree species. Even charcoal made from cedar has been used as pigment, mixed with salmon eggs to make a paint sometimes applied to paddles and other items. Mixed with elk fat, this charcoal has also been used to produce face paint. These are just a few examples; describing all of the uses of cedar wood by the ancestral Quinault could fill an entire book.

Cedar Bark

Just as cedar has provided Quinault with the tribe's most important wood, cedar bark has been the single most important traditional fiber. Bark is pulled off in long strips from living trees. The brittle outer portion of this bark is then removed, leaving only the pliable reddish-orange inner strips, which are soaked and pounded with a wood or bone bark-beater to loosen the strong bark fibers. When pounded long or hard enough, the bark fibers separate and become cottony soft. Prepared this way, cedar bark is then used loose as an absorbent material, or woven and braided into a diverse assortment of traditional goods. Like many other parts of the tree, the bark was historically used from birth to death: cedar bark was used to make baby's diapers, for example, as well as baby blankets and soft mattress bedding. A pad of cedar bark was traditionally

"First thing I'd tell [people using cedar bark] is that you have to soak it for a long time so it's really wet, all the way through. And the wetter it is and the quicker you do it [the easier it is to use]. I've got some that's been sitting, so it's going to be tough. You just get some sharp knives and start peeling that bark off and you peel it again and you have to work very carefully. I don't know, every once in a while I make a mistake and end up with little pieces: little baskets. Then, I usually kind of roll it up and keep it that way and when I want to use it, I put it back into the big barrel and [it] gets damp again and I can start using it. . . . If you peel that [outer] bark off right away, you're ahead of the game because it will come off easier. I have some now that was given to me and it's got the [outer] bark on it and everything and . . . I'm going to be sitting on my new deck, just working on it!"
—Marlene Hanson

"There were several reasons why our people settled out here. In a place like this, [there is] all the necessities for our way of life, the things that were used for anything that was culturally significant. This cedar tree was one of the biggest reasons why they occupied this area. There were two things that were really attractive . . . the salmon and the cedar tree. They had everything here—everything they ever wanted for their quality of life. They lived on this land with what was here and along came the white man and ruined it all. Took all of the cedar trees away."
—Chris Morganroth

placed against the heads of infants and tightened with cedar bark straps to give infant heads a noble, flattened look; and bundles of cedar bark were sometimes taken by guests as a gift to newborn children. On the opposite end of life's arc, cedar bark was widely used to wrap the bodies of the dead for everyday burials and for ceremonial treatments of human remains.

The flexible and versatile strips of inner cedar bark have also been widely used in making baskets of many types. Some cedar bark baskets are made with such a tight weave that, when soaked slightly, they become waterproof and suitable for carrying water or cooking food. Another type of traditional cedar bark basket is made to be sealable, with a woven cedar bark lid—this type is sometimes used to store meat, fish, or berries, and crafted as a "checkerboard" wrapped weave. Cedar bark has also served to form the foundation for baskets made of other materials such as spruce root, cattail, and beargrass. Woven or braided, cedar bark is traditionally formed into all manner of clothing: loose bark skirts for women, rain capes, conical hats to keep off the sun and rain, leggings to protect the legs when walking through brush, shirts, robes, dresses, headbands, armbands, and the like. During the intense shinny games of long ago, players were said to wear only cedar bark headbands—the Quinault always dyeing theirs red, the Queets dyeing theirs black.

Cedar bark was used for many other purposes as well. Woven, it was made into a variety of mats used for floors and wall partitions, and for tents and sails for canoes. Pounded and woven cedar bark has been used for towels and napkins. Handbags made from cedar bark can be used for carrying such items as fire-starting drills, gambling gear, and beads. Cedar bark was also popular for making "tumplines," allowing people to support heavy burden baskets with their foreheads. With several strands wound together, the bark can be fashioned into string and rope, slings and handles, light nets and fishing line; and temporary ropes are sometimes made by peeling and winding together long strips of bark from cedar branches. For heavy loads, even whole spring cedar branches might be twisted together to form tough, temporary cables. Dolls have been made from bundles of cedar bark, tied off to

Quinault museum director, Lani Chubby, processing freshly peeled cedar bark for later use.

make the head and limbs, and outfitted with tiny cedar bark clothes. Cedar bark flags have been used for marking territorial boundaries or items claimed on the beach. Traditionally understood to have ceremonial and medicinal properties, loose or woven cedar bark has been a principal material for headbands and ceremonial regalia. So too, cedar bark has sometimes been used instead of buckskin on drumstick heads. Historically, at the onset of puberty, girls' faces were rubbed with cedar bark towels to prevent future wrinkling.

Many traditional uses of cedar bark faded somewhat with the introduction of Hudson's Bay Company blankets and trade cloth in the nineteenth century, but cedar bark has continued to be used for special occasions to this day. Cedar bark is used widely for the creation of traditional baskets, hats, ceremonial regalia, and sewed fringes on both ceremonial and everyday clothing. As cedar bark and cedar bark baskets were once important trade goods between tribes, they are still actively traded and shared between traditional weavers and others today.

Cedar is often used in building fires. Not only has cedar been used for firewood and kindling, but fires are traditionally lit with cedar bark tinder and a portable cedar torch or punk stick. To make the punk stick, Quinault ancestors used tightly braided or twined bark; bound tightly, a six-foot length could easily last all day. These were sometimes lit and carried from villages to campsites

"You don't want to take the bark off 100 percent, unless you're trying to kill the tree. We don't want to kill our trees. I mean, my gosh, it would be terrible if all of that was gone. The value. We want the trees there for future generations and we want them. We want to protect the trees."
—Marlene Hanson

"The old ladies used to make water sacks out of cedar bark. Real tight. So it would hold water. And they would also boil water, just like a paper cup. But this way, when they took a canoe from Queets to Hoh or from Hoh to La Push, they could have fresh water with them."
—Conrad Williams

so that travelers did not have to light a new fire upon arrival.

The gathering of cedar bark has followed certain patterns since time immemorial. Traditionally, bark gatherers assemble a small camp at suitable cedar groves, with tents made from hemlock bark or cedar mats. Historically, men were often involved more in bark peeling and women more in bark processing, though such divisions varied. Approaching the tree and offering respects, harvesters make a horizontal cut in the bark close to knee height—a cut of perhaps twelve to eighteen inches in width on big trees, and just a hand's width on smaller trees. (Some say the smaller trees produce better bark, while others prefer the older trees with their thicker bark and longer limb-free segments.) Because long pieces of bark are preferred, harvesters seek trees with few lower limbs; and trees with abundant cones are sometimes said to be best. No matter what kind of tree they peel, harvesters say they never take more than one or two strips, totaling roughly one-quarter of the tree's circumference. This was formerly accomplished with specialized bark-cutting tools made from bone, horn, or hardwoods like yew, but more recently with hatchets, saws, or other metal tools. Grasping the bark tightly in their hands, harvesters pulled bark from the tree's side, sometimes twenty feet or more up the trunk, the bark strip tapering the higher up it gets. Harvesters peel as high as they can, then grab the bark over their shoulder and turn with a jerk to one side, popping the top end of the stripped bark from the tree. Long pieces can be hard to break off, and some tribal members describe having to swing from long pieces or even tie them to a car bumper and drive away to remove pieces too long to pull by hand.

Once peeled, the bark is spread on the ground with the inner bark facing upward. As quickly as possible, the rough outer bark is broken off and peeled from the soft inner bark. If the layers are not separated within a day or two of gathering, the inner bark will become almost impossible to remove. Thus, separation is usually undertaken immediately at the tree-side so the outer bark can be left in the forest. The long pieces of inner bark are then cleaned and soaked in water, and the strips hung briefly outside to dry, before being brought indoors for final drying to avoid mold and mildew accumulation. After a couple of days, the strips are dry enough to be folded or rolled for long-term storage. When cedar bark is needed for cultural uses, lengths of the bark are dampened and cut into strips with dimensions suitable to the task at hand, sometimes by pulling the long strips of bark over the edge of a firmly secured yew wood paddle or other wooden bark-splitter. If the bark is to be used for clothing or another soft item, it is traditionally pounded with a bark-beater tool—a tool once made of hard bone taken from a whale's skull. Pounded sufficiently, the fibers separate and become almost as soft as modern cloth. To make it even softer, the bark can then be rubbed by hand. Lightly colored when first peeled, this bark turns a light reddish-brown when exposed to light. It is often dyed with red alder bark, making the cedar bark even more brilliantly reddish-orange. People also sometimes dye cedar bark black, for use in specialized items such as regalia, soaking the bark with various substances such as rusted iron. Elders note that the best cedar bark weaving was done in winter, when there were relatively few outdoor duties; thus they stockpiled cedar bark in the spring in anticipation of the weaving tasks ahead.

Peeled inner strips of cedar bark, being processed for later use.

Other Uses of Cedar

Like cedar bark and wood, cedar roots are of great traditional importance. These roots have been used to produce baskets, hats, mats, and other woven items. Being harder and tougher than cedar bark, the roots make a classic type of "hard basket"—watertight for cooking, and durable enough for gathering berries year after year. Cedar roots have also been used to make strong twine and even fishing nets. Harvesters usually seek roots of one or two inches in diameter and three or four feet long, gathering them where roots are long and straight, such as in loose, sandy soils or from old rotting nurse logs. As with the selective

"The old Indians, when they used to go up to do canoes, they might find a 150 foot tree or something, they would cut, or make a fire about this wide around the tree, not a flame, either, just hot coals, keep it going all the way around, until the tree finally fell. If it came down on you, you sacrificed yourself, that's all." —Conrad Williams

harvesting of bark or plank wood, only a few roots are taken from any individual cedar to avoid harming or killing the tree. Harvesters remove the root's outer bark, split the root into long thin pieces, soak these strips, then dry and save them for later use. The shiny outer part of the strips is used for visible applications, such as the outer weave of a basket, while the dull inner part of the root is used for interior work, such as a basket's interior structural form.

Even tough and pliable cedar limbs were valued for use in traditional manufacturing. When split, these branches were often woven together with split cedar roots to make open-weave burden baskets, suitable for carrying heavy loads of firewood or shellfish. Split into thin strips, the branches were also used to produce watertight baskets. If hunters killed an elk or some other animal far from home, they sometimes made temporary meat-carrying packs from interlaced cedar limbs and bark. Depending on the design, meat piled on these structures could be carried by one person, or two people in tandem. Similar devices were constructed as stretchers for people injured or fatigued far from home; moreover, tied between trees, cedar branch slings served as hammocks. Limbs, stripped of foliage and soaked in water, could be twisted into strong, rough ropes that were sometimes used as handles for baskets or dipnets. Thin cedar branches were also wound together to make line for whaling harpoons. These ropes were tied to sealskin floats so the whale could not dive, and were used to pull whales into canoes or onto shore. They could also be used to tie logs together, constructing rafts for river or bay use. Cedar foliage has long been important for ceremonial purposes as well—used for cleansing in preparation for rituals, dangerous work, or death. Furthermore, the greenery was used to remove both smells and disruptive powers from hunting and fishing gear.

Elders attest that cedar trees are traditionally deemed sacred. The spirit of a tree is said to live on, in a way, within the canoes and houses containing their wood. Ceremonial regalia have always included the wood and bark of cedar; and historically, shamans' highly specialized clothing and tool kits were made of cedar. Cedars are

Elders continue to teach their children and grandchildren the methods and the ethics of traditional cedar bark harvest.

said not only to deliver material goods, but also to give songs and strength to those who engage with the trees respectfully, taking cedar materials only sparingly.

In recent times, many Quinault men have worked in cedar logging and milling, or made shakes and other items from salvaged wood taken from fallen logs in good condition. Many developments—from industrial logging to land privatization and park creation—have created new challenges when accessing cedars for cultural purposes. In response, both Indian and non-Indian loggers have sometimes brought cedar bark and other discarded cedar products to tribal members, or made bark available at industrial log-sorting yards. Gatherers increasingly attempt to coordinate with logging operators to access areas for traditional harvests of bark and other materials prior to clear-cutting.

Beyond the uses described above, cedar is used for a range of everyday medicinal purposes. A number of tribal members report making effective diabetes remedies with decoctions of cedar cones (green cones are said to be best) or immature male cones ("cedar berries"). Roughly a handful or two of cones are simmered in a gallon of water, with a small amount of this decoction being consumed daily to combat symptoms of diabetes. The bright-green new growth on the foliage is also simmered, sometimes with its cones and seeds, to make a decoction to alleviate cold and flu symptoms, or to support general cleansing.

"It's fun to [peel bark], but you got to have strong women or strong men . . . . Some people use different tools. When I go I have a hickory knife [and] I get it started on the sides and the bottom. Some people use a little hatchet. And you get that going and you have to give it jerks and twists and the best thing is to have a man that's not afraid of work come out there with a length of rope that's willing to fall down and get laughed at and snapping it off because there's nothing more embarrassing than to get this really nice piece of cedar way up there twenty to thirty feet and you can't [get it]. . . . You give it a twist and a big jerk."—Alicia Figg

Some families make a decoction of bark and twigs for kidney troubles. Cedar decoctions are also sometimes used to wash sores, and cedar foliage is used in poultices. The roots were also said to have been used medicinally in a manner similar to the greens, especially in making decoctions and teas. Historically, Quinault ancestors wrapped tufts of cedar bark around a certain plant from the mountains, rolling this into a tight ball and igniting it on top of sore, rheumatic joints for relief. Called "moxabustion," this was a practice widely known in Chinese and occasionally in Native American medicinal traditions.

Another cedar of enduring significance to Quinault people is the Alaska yellow cedar (*Callitropsis nootkatensis*). Found especially in higher-elevation forests, tribal members have sometimes sought out the prized wood of this tree in the Olympic Mountains and other places in the region. The wood is remarkably dense, durable and rot resistant, especially in older trees from higher elevations. The tree's wood is characterized by its straight grain, its waxy and workable texture, and uniform light-yellow color that gives carved items an attractive finish. For these reasons, carvers consider the tree a relatively rare but important specialty item –sought by tribal members for specialized use in the carving of paddles, carved ceremonial items, bowls, utensils, and certain traditional wooden tools. The dense wood of the yellow cedar also burns hotter and longer than many other tree species, too, making the tree a popular source of firewood. Because this tree is so slow to rot, harvesters can sometimes recover wood from dead trees for carving or firewood, even decades after the tree has fallen. The tree is widespread in the Cascade Range, as well as along the coasts of Alaska and British Columbia. Being relatively hard to find in accessible and harvestable stands, some Quinault families return and sparingly harvest from certain small patches of trees over generations, or obtain yellow cedar wood from family and friends in other tribal lands, where the tree is more abundant.

What It Looks Like

Western redcedar is a towering tree, often surpassing 200 feet, with a broad, fluted, and often buttressed base. Glossy green foliage, scaly and distinctive, hangs from drooping branches that bear numerous tiny reddish-brown cones. The fibrous bark, ranging from gray to reddish-brown, runs vertically up the tree and is easily peeled into strips. Ancient trees often have a "candelabra" top, consisting of multiple gray spires—from many years of treetops dying and being functionally replaced by upper branches.

Alaska yellow cedar can be told apart by its far droopier branches, lacking the upright tip of a J-shaped western redcedar branch; when the foliage of yellow cedar branches is crushed, it does not emit the pleasant and familiar smell of redcedar, but a vaguely unpleasant smell that some compare to mildew. Distinctive cyprus seed cones also set the tree apart—relatively spherical, green and compact at first, becoming open and woody-brown with four to six woody scales upon maturity.

The signature buttressing and reddish-brown bark of a young cedar tree.

Where to Find It

Redcedars are found widely, though unevenly distributed, in well-established coastal forests. The tree is best sought a short distance inland, as cedars often struggle from wind-throw and salt spray near the sea. Traditional gathering areas were found in many locations, including near the margins of anthropogenic prairies.

Historically, redcedar logs for canoes and large planks were often gathered in special places close to the ocean or on the lower rivers, in part so the wood could be easily floated to villages for finishing; in recent years, because of logging and other development, few old cedars remain in these areas. Cedars growing at higher elevations in the Olympic Mountains have tighter-grained wood, very desirable for canoes and other use, but it was often difficult to transport such logs to navigable water below, and such trees are now usually off-limits to harvesters within Olympic National Park and other protected lands.

In contrast to the redcedar, Alaska yellow cedar is found at middle to high-elevation forests in Washington, primarily in the Olympic Mountains and Cascade Range. Outlying, often isolated groves can be found in other locations. The tree is common on the British Columbia Coast, and especially along portions of southeast Alaska, replaces redcedar to become the dominant cedar of the coast.

When to Gather

Cedar bark is peeled when the sap starts to run in the spring, roughly in April. Traditionally, this was often done during trips inland to dig camas and other roots. By summer the bark hardens and can no longer be easily pulled, though larger trees can sometimes be harvested later—often well into the summer. Some elders reported that the bark was traditionally gathered between the time the geese flew north and when they began flying back south; others say that the peeling season began when the salmonberry blossoms were well along, and stopped when the weather got hot. Some elders report that the bark is often hard to peel after long periods of wet weather; in part for this reason, bark is usually gathered during sunny weather. Infrequently, people harvest cedar bark in the winter months, when it is needed for regalia or other purposes, but this is done with difficulty. Wood and limbs can be gathered anytime. Roots are often gathered in winter, though some harvesters choose to gather in late summer, when the new roots are strong but still very flexible.

Traditional Management and Care

Few species are the focus of as much traditional and ceremonial respect as western redcedar. Many accounts suggest that harvesters of cedar bark, roots, or even whole trees traditionally show respects to the tree. The spirit of the tree is traditionally understood to not only be conscious of harvesters' actions, but to live on in myriad ways within the objects made from the tree. Respectful engagement with the tree is thus essential if a canoe or house is to be safe, or if regalia or medicines are to have their desired and positive effects. This also contributes to a range of conservation practices. Elders note that only a small quantity of cedar bark or roots would be gathered from a tree to avoid harming or even killing the tree; apologies or offerings might be given during the process. Many of these values and traditions endure.

Modern harvesters continue to observe many traditional protocols, but also to navigate a world in which this important tree is treated as an economic commodity. Tribal harvesters increasingly try to coordinate with logging operations, on reservation and off, to get access to cedar trees so that they can peel bark and show respects shortly before an area is logged. Industrial logging has made old trees relatively scarce, so large logs are used sparingly and only with serious intent. Poachers on the reservation have taken cedar foliage for floral industry use and have sometimes felled entire mature trees to strip them clean; they also take blocks of wood from downed trees. Quinault Indian Nation laws prohibit these activities, and the tribe enforces against illegal harvest—not only because of the economic loss to the tribe, but because of the enduring cultural value of the trees.

Cautions

Some of the same essential oils and compounds that make cedar aromatic and rot-resistant can cause skin, respiratory, and digestive irritations in some people—especially after long or heavy exposures. Among these is the hydrocarbon called "thujone," named in reference to the scientific name of the tree, *Thuja*. Anyone who inhales or ingests large quantities of cedarwood powder, cedar foliage derivatives, or other pieces of the tree risks lung problems, digestive upset, or even heart and nervous system trouble. A small number of people report negative reactions to cedar cone tea; those unfamiliar with this beverage should start with cautious small doses. People who pull or work with cedar bark extensively, or expose the bark to open wounds, can get "cedar bark poisoning"—a range of conditions from skin infections to flu-like symptoms that have sometimes been remedied in Quinault country by consuming repeat doses of crabapple tea.

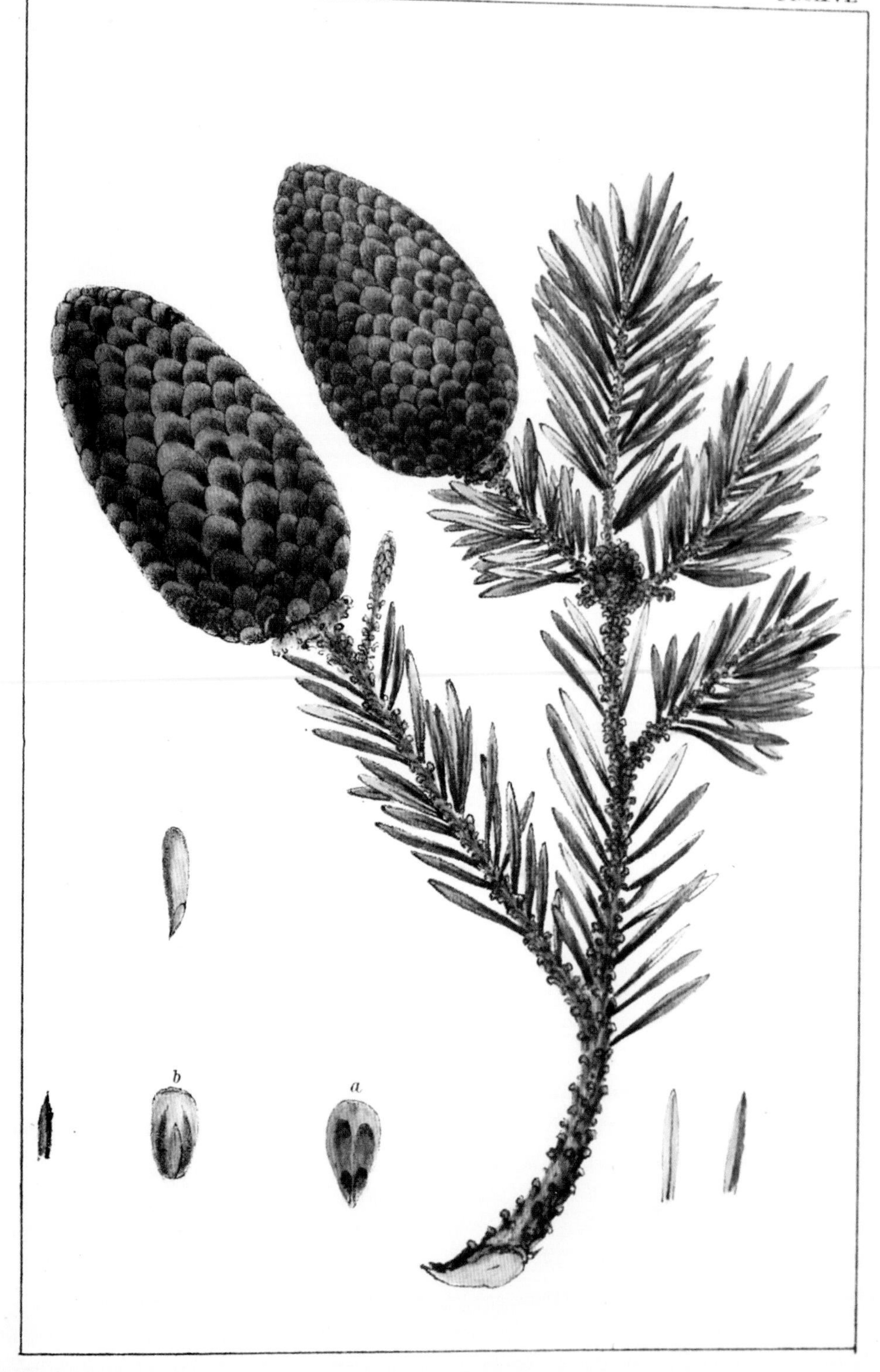
Pl. CXVI.
b
a

Sitka Spruce

kwaʔa lel | *Picea sitchensis*

Traditional Uses

Sitka spruce is among the most useful species traditionally used by Quinault people, while being, at times, underappreciated by the wider world. The lightweight, strong wood is uniquely adapted to the windy ocean coast of the Northwest, with its long shallow roots allowing spruce to remain upright in saturated coastal soils. The roots and broad lateral branches are also strong. They bend but are hard to break. In fact, Sitka spruce has some of the strongest wood per unit of weight of any tree in the forest. It is so strong and lightweight that it was once a primary source of material for manufacturing airplanes, such as the World War II-era plane Spruce Goose. In fact, some of the earliest industrial logging in Quinault country was undertaken during World War I, delivering Sitka spruce to the military for the construction of fighter planes.

The giant columnar trunk of an ancient Sitka spruce, with mosaic bark of gray scales, lifts its prickly, needled branches high above the forest canopy.

The toughness of the wood meant that fine woodwork with it was not always possible. Though small dugout canoes were occasionally made from Sitka spruce, and though the canoes were extremely durable, they were not as elegant as cedar canoes. Strong house planks—removed from logs or living trees with wedges, mauls, and adzes—were sometimes made from Sitka spruce as well, as were house frame poles. Spruce was used for handles and poles of dipnets, and peg-like spruce wood dowels, or thongs of spruce root or limb, were traditionally used to fasten pieces of wood when making bentwood boxes, canoes, or other large wooden items—running the cord or dowel through a drilled hole so that it functioned like a screw or nail. The strength of spruce wood also made it a popular material for digging sticks

"Spruce roots were dug during the winter. Roots from pencil size to three inches in diameter were made use of. They were cut into four-foot lengths and the bark removed by holding them over the fire for a few minutes, then peeling. They were then split, beginning at the small end, into quarters, sixths or eighths, with mallet and chisel. These sections were then split into layers about one-sixteenth of an inch in thickness by holding one piece in the teeth and stripping or pulling off the thin piece with the right hand, the left following down the section at the point of the split. The roots split fairly readily between the rings, provided the process is begun at the point of the triangular piece [the center of the root]. These pieces of varying width were then reduced to the desired width by further splitting. For watertight baskets the fibers must be almost square [i.e., about one-sixteenth by one-sixteenth or slightly larger]. The strands were then soaked in warm water for an hour or so, then scraped by holding a shell in the hand and pulling the fiber between the edge of the shell and the thumb. The fiber was then stored away, or stored before scraping. In either case it was soaked before using. This material was used especially for hats, water baskets, berry baskets, or other waterproof containers and also for openwork carrying baskets."—Ronald Olson, 1936, based on the accounts of Quinault elders

for clams and roots—one common type of digging stick was roughly three feet in length and made from spruce limbs or the extremely tough spikes of limb heartwood found inside large, rotting spruce logs. In addition, this interior limb spike, often long and conical in shape, was sometimes used to make torches, with embers placed in the natural depression found on the thick butt end. Dried pitchy chunks of spruce wood have always been popular as fire-starters. These are mixed with damp firewood to start a roaring fire when dry wood is scarce.

The roots of Sitka spruce have been especially important in Quinault traditional manufacturing. Sitka spruce roots are hugely important in basketry and other weaving traditions, as is true wherever this tree is found—from the northernmost California coast to southeast Alaska. Within the State of Washington, Quinault traditionally use more spruce roots for basketry than most other tribes. Traditionally, the strong roots were ideal for baskets meant to carry heavy loads. These included burden baskets for carrying large quantities of fish, clams, or berries—a kind of "openwork" basket allowing contents to be washed by dunking the basket in rivers and streams. The openness of the weave was determined by the intended use of the basket. Yet, spruce is also used for many other types of baskets, each crafted with distinctive techniques, from coiling to weaving to twining. Many are considered "hard baskets," as opposed to softer grass-based baskets, with split spruce roots used for the interior structure of the basket as well as the weaving. These spruce root baskets are woven tightly and swell a little when moist, making them watertight. Thus, they have often been used for holding water and for cooking. Historically, many of the simmered food and medicinal preparations described in this book were made by placing fire-heated stones in a spruce root basket full of water. Sitka spruce is also widely used as the structural foundation for baskets that feature beargrass, cattail, sweetgrass, and other showier materials on their exterior, though spruce root, with its deep brown color, is attractive enough to be included on the exterior of baskets, often decorated with white beargrass leaves, black maidenhair fern stems, or other materials of contrasting colors.

Beyond their use in basketry, spruce roots have been twined or woven into other traditional products. The same properties that make spruce roots ideal for the production of watertight baskets make them useful for rain gear. Traditionally, spruce roots were fashioned into rain capes and conical hats. In these garments, the spruce was sometimes used alone, but more often as the core structure through which cedar bark or other materials were woven. Sitka spruce roots were also woven into mats. They were especially popular for making durable mats for heavy uses that might damage softer materials. Often twined into rope, cords, or string, spruce roots were traditionally used for basket handles, or to manufacture certain types of fishing gear; for example, fishermen traditionally used Sitka spruce string to tie the tines of salmon spears to the wooden spear shaft.

The roots are often sought in soft ground, such as sandy places along beaches, riverbanks, and bays, where they typically grow long and straight, and can be easily dug from the earth. In rocky areas, or in deep forest soils, the roots are often gnarled and complexly branched. Some harvesters gather spruce roots that are growing on decaying "nurse" logs. Some of the very best root gathering in Quinault country is said to be among the freshly fallen trees and historically vast logjams along the Quinault River and other waterways. Harvested roots range from roughly pencil width to nearly three inches in diameter. These are cut into lengths, and the exterior is charred quickly and carefully to remove the roots' bark. If the roots are heated too intensely or too long, their color and flexibility changes. Weavers usually split the roots into long thin strips—often splitting each length into quarters, sixths, or eights before soaking the strips in warm water. Any soft interior wood is then scraped off, leaving only thin, mostly flat, durable strips of root. These strips are

"*The pitch of the spruce was used for a lot of things. We would mix it with oil, like bear oil, and they would use it as a paste. Whenever they found their canoe cracking or something like that. It had to be a real fresh—otherwise it was kind of crunchy. . . . We would have to melt it. And they also used it as a paint because they could burn it and use the black powder. They pulverized it and mixed it with beargrass, I mean bear grease or whale oil, any kind of oil that they had. . . . The reason they liked it . . . was because it didn't reflect. And if you had, like, a glossy paint it would reflect the sunlight and scare your animals away.*"—Chris Morganroth stored, then resoaked for softening as weavers prepare to make spruce root baskets or other items.

The young, springy limbs can be wound together to make a thick rope, including rope so strong it was historically used to tie whaling harpoons to canoes, and to haul whales to shore. Additionally, spruce limb ropes were used to tie large house planks in place, tethering them to the house frame and to each other.

"Spruce pitch: [we] put it on fish spears so you wouldn't have any reflection . . . and we used to pick spruce roots on up the Quinault . . . on the river bottom, wherever they've uprooted."
—Justine James Sr.

Spruce pitch is also one of the more important natural multipurpose substances available in Quinault country—widely used in the production of traditional paints, adhesives, and sealants. Heated slightly, spruce pitch can be used for caulking canoes, and paddlers often use spruce pitch for quick repairs of cracked canoes. It is also used to seal the seams of bentwood boxes and other watertight wooden items. If spruce wood or pitch is heated, it produces a black powder that can be mixed with other materials to make a matte black finish. Elders such as Chris Morganroth note that the blackening was sought because shiny finishes on canoes, fish spears, and other items "would reflect the sunlight and scare your animals away."

Yet spruce pitch is also a keystone traditional medicine, having antibiotic properties, and is used as a salve and sealant for cuts, burns, and other wounds. Formerly mixed with bear grease to make a salve, the pitch is increasingly mixed with beeswax, petroleum gel, and other substances to reduce stickiness. Some families still make salves for personal use, as gifts, or occasionally for commercial sale. Spruce pitch has also been used to draw infections and other internal ailments from the body, often in combination with other ingredients. The pitch can be chewed like gum, usually after it has been slightly warmed; some elders report this to be nutritious and healing. A tea made from pitch or pitchy inner bark is traditionally made to soothe an irritated throat. Along much of the coast, pitch has been gathered where it is naturally available, though people sometimes cut the trees, placing a board or other object to capture the pitch below; some Quinault have kept "pitch wells"—carved, concave notches in the sides of trees that capture pitch draining from the cut bark above. Once the pitch has been gathered, people often warm the pitch and filter it to remove impurities, then mix it with other ingredients before cooling and storing it for later use.

Sitka spruce has important health benefits beyond the value of its sap or pitch. A decoction made from fresh spring spruce buds was once a popular tea, used by some families to this day. This was sometimes brewed stronger for specific medicinal uses, including colds and cleansing. A decoction of spruce limb bark has been consumed for the symptoms of cold and flu. A decoction from spruce buds or sap has also been used as bathwater, widely reported to be strengthening and invigorating. Some elders recalled eating the new spring needle buds when they are soft and brilliant green—a nutritious and healthy snack, rich in Vitamin C. Some ate only the innermost soft stem, peeling off the soft new needles, which can become mealy. These spruce buds can serve as an appetite suppressant, used when traveling or when fasting for ritual purposes such as during adolescent vision quests. Quinault people also traditionally use Sitka spruce boughs widely in ritual contexts, such as the scrubbing of the body with spruce boughs to cleanse before ceremonies. Occasionally, people's remains were even placed in spruce trees after death. The tree has a rich and important role in oral traditions, reflecting the overall cosmological importance of spruce trees within Quinault custom.

What It Looks Like

Spruce is a broad, fast-growing tree with strong horizontal branches, and one of the world's tallest trees, commonly surpassing 200 feet in height. Branchlets are spiky in younger trees, drooping on older ones, and are dotted with small nubs at the needle attachments; branchlets bear needles that are stiff and sharply pointed. The pendulous, light-brown cones, with numerous thin scales, are roughly 2–4 inches long. Scaly gray to reddish-brown bark breaks off in large puzzle-shaped pieces.

Where to Find It

Sitka spruce is especially abundant near oceans and bays; along much of the Pacific Northwest this tree is seldom found beyond roughly fifteen miles inland from salt water, but also extends much farther up low-elevation, well-watered river valleys such as the Hoh, the Queets, and the

Quinault. Roots are best gathered in areas with sandy or loose soil, such as along sandy beaches, riverbanks, and bays—but also in places with loose glacial soil. People traditionally look for uprooted trees, or places where erosion has exposed roots in riverbanks and beaches, to minimize adverse effects on living trees.

"Spruce roots I gather close to the beaches because the sandy loam—it makes the roots grow longer. I actually did the cooking . . . over the fire. After doing it a couple times, that outer bark comes right off. [In] the sandy loam, they're nice and straight [and] the roots will follow right along the nursing logs if they're well-rotted."
—Harvest Moon

When to Gather

Spruce wood and limbs can be gathered anytime. Spruce buds and sap are gathered in the spring, with sap-gathering often continuing into the summer. Winter is often said to be the traditional time to gather spruce roots in Quinault country; still, some people have gathered in early summer, when the newly grown roots are strong but flexible.

Traditional Management and Care

Root-gatherers take only a few roots from each spruce, attempting to not kill or seriously harm the tree. Pitch-gathering is also done respectfully, cutting into a tree only when needed and not in such a way as to undermine the tree's health. Experienced harvesters often show great respect when gathering pitch, roots, or other material from living trees.

A traditional spruce root basket, one of many on display at the Quinault Cultural Center and Museum.

Bentwood cedar boxes, large or small, are held together with spruce pegs passing through drilled box corner holes.

Springtime brings newly sprouted spruce cones and edible buds of soft spruce needles.

Spruce dominates the forests fronting the open ocean, where salt spray and high winds keep other tree species at bay.

Western Hemlock

wapɫ | *Tsuga heterophylla*

Traditional Uses

In Quinault country, western hemlock is commonplace. In forests that sit undisturbed by fire or wind for long periods, hemlock becomes dominant—towering with its furrowed gray trunk and lacy green foliage over the dark forest below. Even in the darkest forests, hemlock seedlings thrive on stumps and downed logs. The mountain hemlock (*Tsuga mertensiana*), similar in appearance to its lowland relatives, can be found high in the Olympic Mountains and is used for many of the purposes described below.

Though its wood is softer and more subject to rot than that of other conifers, hemlock still plays an important role in Quinault cultural practice, especially for uses requiring temporary or biodegradable wood. Hemlock wood is commonly used for firewood, but it is also used for other purposes. When Quinault ancestors built temporary defensive palisades around their villages in times of war, the strong main posts were cedar or cottonwood, lashed together with crossbars of young hemlock averaging four inches in diameter. The wood was also carved into utilitarian items like combs, as hemlock wood is easy to carve into fine-grained products not necessarily durable or made for long-term use. Hemlock sticks or saplings were also used to build fish traps, used as structural poles, or stuck into the river bottom as the "wattle" sticks through which more flexible materials like cedar limbs were woven. Moreover, people have used hemlock poles to build simple walkways over fish traps and stations. Here too, the wood was often used (and even valued) as a biodegradable, disposable item, replaced each year after winter-high water had damaged the prior year's traps. Young hemlock trees are sometimes cut and used for all-purpose poles, for such uses as canoe punting poles of roughly ten to fourteen feet, used for traveling upstream. These poles were cured carefully, with the butt end down and the tips hardened by light charring in fire. Easy-to-find hemlock saplings, relatively strong and flexible, were often lashed together to make dipnet hoops as well. And hemlock was used to construct sports equipment, with some people fashioning small spears that were thrown at bracket fungus targets in point-keeping competitions. So too, people sought hemlock sticks, especially those with a natural curve on one end, to use in shinny games—the ideal stick length being from the middle of the player's chest to the tip of the middle finger.

"I used to trap for years until I got too old for crawling around in the woods. . . . I used hemlock bark. Hemlock bark will turn your trap real black. Takes the shine off of it. [The trap] has a little shine to it, and it's just enough that your prey or predator will see it, before they'll step on it. . . . Fact is, I think they used it to paint clothing and things [with tree bark] at one time."
—Francis Rosander

Hemlock bark, like the wood, was also used for a number of temporary purposes. Temporary shelters were sometimes made of bark from larger hemlocks, either salvaged from dead trees or removed from living ones. This was especially done during inland springtime root-gathering trips, when the trees' sap was running and the bark was easily removed from live trees. Hunters and other travelers made temporary shelters of poles covered in layers of the bark. If thick enough, this layering insulated the interior, making it nearly waterproof. Thick pieces of hemlock bark were also used to construct temporary dwellings at villages, especially when visitors were too numerous or dodgy to bring into peoples' homes. These structures were built with a frame of small poles resembling that of a longhouse, but smaller. The use of hemlock bark over cedar reflected, in part, the great value and difficulty of obtaining good cedar planks.

Hemlock bark was also central to aspects of traditional Quinault food cooking and storage. It could be soaked in water to soften it, then bent as needed and stitched in place to make loose, boxlike containers for storing elderberries and other foods. One such traditional Quinault food storage box is constructed by placing two strips of the bark crosswise in a cross shape; the sides are then folded up over the overlapping square, creating one-ply sides and a two-ply base that are stitched together in the corners. The berries placed inside these containers were usually covered in leaves. Historically, hemlock containers were submerged in streams or springs to preserve the food for

The bright green needles of hemlock are short, lacy, and soft to the touch.

the longer term. Stored by this method, berries could easily last through winter. These containers could also be used aboveground to store foods, from meat to berries, by filling the containers with oil that would solidify as it cooled, sealing its contents and making them airtight. Such hemlock bark containers were also used to line cooking pits and as cooking containers. To do this, people traditionally placed large strips of bark so that they lined an excavated pit. Hot water was then poured over the bark so it became flexible and conformed to the shape of the pit. Once cooled, this bark shape was removed and stitched together, ensuring a pit-perfect fit. The bark liner could then be filled with water and hot stones, for example, to boil foods or even simmer medicinal plants within the pit. The Quinault have ancient oral traditions about peeling bark from hemlock

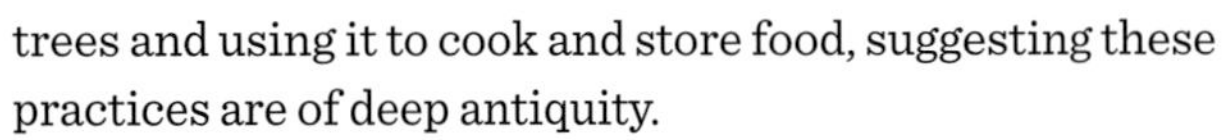

trees and using it to cook and store food, suggesting these practices are of deep antiquity.

"There are certain things you have to do, and some of these things involve hemlock and spruce and [cleaning] yourself with those boughs."
—Guy Capoeman

Meanwhile, the soft inner bark of hemlock was widely used as a dye. This inner bark was simmered and sometimes mixed with salmon eggs to make a stain for nets, dipnets, and paddles—in part to make the objects less "bright" and visible to fish and game. When Quinault people adopted metal traps, they began to boil these with hemlock bark to eliminate their metallic shine. The duration of boiling and the amount of bark controlled the color and darkness, which ranged from yellow-orange to deep, dull brown. Knowing this, some basketmakers simmered hemlock bark to produce dye, especially to lend a brown color to entire baskets or basket design work. The inner bark was probably also eaten, as is common in other tribal communities along the coast.

Western hemlock had a range of everyday medicinal and cosmetic uses as well. A decoction of boiled bark is sometimes used as a laxative. Furthermore, the bark has been crushed and applied to the skin for venereal diseases. Before tobacco was commonplace, hemlock needles were at times smoked, or mixed with trade tobacco along with such substances as salal leaves. Hemlock trees can exude a dark, almost black, pitch traditionally called *stca'mats*. This black pitch, or ground bark containing it, was mixed with elk tallow and sometimes applied to sick children's chests like a "vapor rub." It was also gathered and used as a perfume for young women. Young men sometimes mixed the pitch with tallow from the marrow of an elk tibia, gifting the substance to young women in a show of affection. After heating the mixture slightly, women would use it as a traditional eyebrow makeup, their eyebrows becoming shiny black with regular use.

Traditional Quinault dip nets and other fishing gear have sometimes been made from the springy branches and trunks of young hemlock trees.

Similar to other large trees in Quinault country, Western hemlock is traditionally understood to have its own unique spiritual significance. Saplings are sometimes braided and used to scrub one's body as part of adolescent training and other activities requiring deep ritual cleansing. Entire boughs and branches are also traditionally used in diverse ceremonial settings, for example, as cleansing to prepare for rituals and other hard work. The bark has sometimes been ground and mixed with pitch and other materials to make a brown body paint, and nets dyed brownish with hemlock bark were said not only to be better (they did not visually startle the fish), but also were "lucky," for reasons likely relating to this quality. Hollowed-out hemlock logs were formerly used in rituals meant to influence the weather and other natural phenomena.

What It Looks Like

Hemlock is an evergreen tree reaching up to 230 feet, with drooping branches lined with soft, feathery, needled branchlets; needles vary in length and are whitish on their undersides. Small, scaled brown cones develop on the ends of branchlets. A telling identification is that the top leader is always drooping over, unlike other similar conifers. It can be readily identified by looking at the top, whether alone or in a heterogeneous stand.

Where to Find It

Hemlock is especially widespread in well-established lowland forests. These trees are sometimes less common close to the ocean or on south-facing slopes, reflecting the tree's susceptibility to wind-throw and its predilection for shady sites. A close relative, mountain hemlock (*Tsuga mertensiana*) can be found high in the Olympic Mountains.

When to Gather

The bark of hemlock is easiest to peel in the spring and summer, as the sap flows, but in lean times can be removed year-round. The gathering of fresh, moist pitch is best done in the spring, but dried pitch can be gathered at other times of the year. Wood can be harvested anytime.

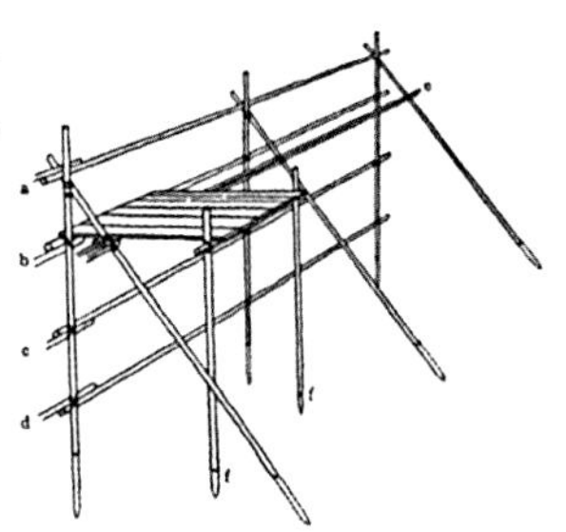

Poles of hemlock, pressed into the beds of rivers and bays, are the foundation of many Quinault fishing weirs.

Traditional Management and Care

As with all trees, hemlock bark and other materials are traditionally gathered selectively, never removing enough to kill living trees. Those who take wood or bark from living trees, especially for medicinal or other cultural purposes, are careful to show proper respects.

Faux-Tsuga de Douglas. Pseudotsuga Douglasii Carr.

Douglas-Fir

ja?mac | *Pseudotsuga menziesii*

Traditional Uses

One of the most common trees on the Olympic Peninsula, Douglas-fir has been especially important to Quinault people for its wood. The densely grained trees in mature forests do not split evenly and are rarely used for carving, yet their wood is valued for its strength. For this reason, Douglas-fir has been prized in Quinault country for the manufacture of arrow shafts, long sea lion harpoons, salmon spears, dipnet handles, and other applications requiring strong, thin pieces of wood. Sometimes these pieces are split from large trees, but they are also sometimes taken from thin understory trees that grow slowly in closed-canopy forests, providing unusually dense-grained wood. Douglas-fir is also traditionally used to make the doweling that holds together wood in the joints of canoes, bentwood boxes, and other traditional woodwork: a hole is drilled in the connected pieces and a Douglas-fir dowel firmly inserted, then cut and sanded so the dowel is flush with adjoining woodwork, before being sealed with pitch-based glue or other natural adhesives. Boiled in water with nets and other fishing gear, Douglas-fir bark can stain items brown, making them less visible to fish.

"My dad made arrows.... The fir made the better arrow if he could find it... they even like that today. There's a certain stage of certain Douglas-fir trees, that their grain is so tight it would dull a chainsaw in just a matter of minutes. That's the nature of the fir tree. It had a really tight grain. That was the most desirable for arrows. It still is today."
—Chris Morganroth

The wood and bark of Douglas-fir are also popular sources of firewood. The thick corky bark—more easily gathered by hand than wood—was especially popular for fires prior to the advent of chainsaws. The bark sheds easily from dead or dying trees. (This fact is mentioned in a Quinault oral tradition regarding two boys, Misp' and his brother, who climb a tall Douglas-fir tree to escape an angry woman, then cause the bark to fall and bury her so they might escape.) Downed Douglas-fir limbs are traditionally stockpiled in a dry place and used as kindling.

If sufficiently dry, this pitchy wood can even be used to ignite damp firewood in winter. Pitch or pitchy bark was likewise used for torches, for both indoor and outdoor use; and long pieces of sticky pitchwood, split three of four times on one end, made an outdoor torch that could stay lit even in the wind.

Douglas-fir has also played an important role in traditional Quinault food and medicine. The pitch is chewed like gum—usually after boiling, as raw or dried sap direct from the tree sticks to teeth. Young needles are simmered to produce a pleasant, calming tea, which some say tastes the way a Christmas tree smells—though darker and older needles are not used in this way. The tea is said to be a popular spring tonic, full of nutrients. Some report that it regulates blood sugar in diabetics and prediabetics. Simmered to a darker decoction, Douglas-fir extract from young needles, needle buds, or roots has sometimes been used in medicines for the symptoms of colds. Douglas-fir pitch, applied to minor wounds independently or mixed with other ingredients, is reported to seal wounds and speed healing.

"On the Douglas-fir . . . the shoots when they first come out, they're nice and light and green, snip those off just a little bit above where it starts to get bright, snip those off, get a handful, just like you would do a coffee, just drop them in and boil them. Makes a great tea. . . . It gives you a calming effect. . . . More like a morning tea, an evening tea or something. Something where you can sit back and enjoy yourself while you're looking around, seeing everything."
—Conrad Williams

What It Looks Like

Douglas-fir is a giant evergreen, sometimes reaching 250 feet, with long, neatly layered, upright branches and medium-length deep-green needles that are strong but soft. Young trees have a triangular shape; older trees have deeply furrowed brown bark. Trees produce clusters of medium-sized reddish-brown cones.

Where to Find It

Douglas-fir is found in low to medium elevations on the western Olympic Peninsula—though it is relatively rare close to the ocean coast. More common in drier, fire-prone parts of the Northwest coast, Douglas-fir is the dominant conifer inland and on the eastern side of the Olympic Peninsula.

When to Gather

Douglas-fir wood and limbs can be gathered anytime. The greenery and pitch are best gathered in the spring.

Reddish-brown Douglas-fir cones, dangling below fragrant-needled branches, have distinctive three-pointed bracts extending from beneath each lobed cone scale.

Traditional Management and Care

Sap-gathering is often done by finding a tree with a natural opening, such as an opening caused by falling trees hitting the bark. When such options are not available, Quinault people have manually cut the bark or even dug out a shallow "pitch well" in the side of the tree that captures dripping pitch in the side of the tree. Harvesters take pitch, needles, and needle buds only in modest quantities, in part so the tree does not suffer or die from the harvest.

Pines

haʔagwal and t̓atnix̣ło (shore and lodgepole pine) taʔckanit (white pine)	*Pinus contorta* (shore and lodgepole pine) *P. monticola* (white pine)

Traditional Uses

Pines grow in two very different forms in Quinault country: Near the coast grows the gnarled and short "shore pine" (the same species as lodgepole pine), stunted in part by its exposure to salty air. Well inland, lodgepole pine and white pine grow relatively straight and tall (lodgepole and shore pine are actually subspecies of the same species). In Quinault tradition, these trees are especially important as medicine. The needle buds and fresh new growth of pines are simmered for use in medicinal teas. Bark and roots are sometimes used in medicinal decoctions as well. Such pine teas and decoctions have long been consumed for general well-being, as well as for respiratory health when people catch colds. White pine bark has also been simmered to make a decoction consumed for stomach complaints or problems involving the health of the blood. This was also said to be a blood purifier, cleansing the entire body of toxins. Reports exist of people with especially bad health issues, such as rheumatoid arthritis, soaking in decoctions that include pine and possibly other ingredients.

Pine sap is also used for medicinal purposes. It can be simmered in tea or chewed like a gum. Sap that has dried and hardened naturally on the tree is not quite as sticky as fresh pitch and is therefore often preferred. The sap or pitchy new needle buds are traditionally chewed for sore throats. Taken from shore pine and its lodgepole relatives from the interior, pitch can also be placed on open sores to seal and accelerate healing.

The use of pine for other purposes is limited compared with other tree species. Shore pine wood, in particular, can be gnarled, pitchy, knotty, and not especially useful for woodwork. The pitchy wood and cones are popular for fire-starter but are often avoided as everyday firewood because pitchy wood generates sparks, as well as heavy smoke that alters the flavor of foods. While not as strong or workable as spruce or cedar roots, pine roots from all locally available species have occasionally been used as a light-duty fiber in baskets. A few basketmakers have even experimented with the use of longer pine needles to make pine needle baskets—a craft learned from tribes outside of the Northwest—but local pines have short needles, yielding limited success. Pine seed kernels are edible and nutrient dense, but are so tiny and labor-intensive to gather that they are seldom consumed.

"Pine tea. I go after that too . . . the little light green parts on the tips . . . . It's good for cleansing your system."
—Micah Masten

The distribution of pines is in flux. Sometimes, pines—lodgepole in particular—have invaded historically burned prairies once burning has ceased. Introduced diseases, such as white pine blister rust, have obliterated whole stands of white pines throughout Western Washington. In recent years, illegal poachers have targeted white pine on the reservation, selling the greens to floral companies that make holiday boughs and wreaths, prompting an expanded law-enforcement presence. Even if the trees are not killed immediately, trees heavily pruned often develop blight and other diseases and eventually die.

What They Look Like

Shore pine is a medium-sized evergreen reaching up to 50 feet, with an irregular, sometimes sprawling shape and prominent branches with medium-length needles, with generally 2 needles per bundle. Its lodgepole relatives are relatively straight and taller—up to roughly 150 feet when mature. Both produce medium-sized egg-shaped cones with hard scales, and needle buds that grow to become tall "candles" of new growth in spring.

White pine is an evergreen tree with neatly layered, sometimes sparse branches and longer, fine needles (up to 5 inches) and have 5 needles per bundle. Dangling cones are thin and long (from 4 inches to nearly 10 inches) and generally curved in a banana shape. Mature trees reach 180 feet and have a narrow, tall, triangular shape.

Where to Find Them

Shore pine is especially found near ocean beaches and clearings nearby. Lodgepole pine can be found in some clearings at lower to middle elevations away from the ocean. White pine can grow at sea level, but is more common inland and on the middle-elevation mountains of the Olympic Peninsula.

When to Gather

Pitch is most often gathered in the spring and summer, when it is flowing within the tree; off-season sap-gathering, while difficult, can sometimes be undertaken in cold and flu season. Needle buds are gathered as they appear in the spring too. Bark and root gathering often occurs in the spring.

Traditional Management and Care

As with other medicinal plants, respects are shown when harvesting. Harvesters are careful not to take so much pitch, bark, roots, or needle buds to harm or kill the tree.

Cautions

Pregnant women should generally avoid pine needle tea, as it can cause miscarriages in a small number of cases.

Especially common inland, white pines are recognized by their long, thin cones and needles.

Gnarled by beachside salt and wind, shore pines have dense curved branches and small compact cones.

XXII, 9.
11. Taxineae
4
5
7
6
2
B
1
3
A
WM
22. Taxus baccata L. Eibe.

Western Yew

ceʔeʔkak or ƛ̓amak | *Taxus brevifolia*

Traditional Uses

A slow-growing tree of the forest shadows, western yew develops a tough, dense-grained wood, lending it unusual weight and strength. Because of these qualities, yew is among the most important trees in the Quinault world, with many different uses.

Yew is traditionally used to craft heavy-duty hunting and fishing gear. It was the standard wood for bows, commonly backed with sinew, and was the preferred wood for arrows that must be especially strong, such as for hunting elk or bear. Yew arrows were often tipped with elk-bone points. Harpoon shafts crafted from yew were used for harvesting salmon, seal, whale, porpoise, and other marine animals—the shafts often eight feet or so in length and tapering at both ends, attached to harpoon points made of bone. The wood's heavy weight helped deliver the harpoon to its target; as Horton Capoeman recalled, "yew had the weight to drive it into the whale." The yew tree was also used to make heavy-duty clubs, such as those used to dispatch seals or sea lions that had been speared and pulled alongside a canoe. Similarly, yew was used to make weapons, including war spears of roughly eight-foot lengths with whalebone or mussel-shell points and, occasionally, clubs and protective armor made of wooden slats. The strong wood also served as the spring mechanism in snare traps for large mammals such as deer, suspending a snare loop made of rawhide. Dipnet frames and handles were made from yew saplings tied together and sometimes steamed. These were roughly oval-shaped, about five by seven feet across. Though other tree species might work for this application, yew was best for hauling heavy fish, or for working in rough conditions that might damage other woods. Furthermore, yew wood digging sticks were standard equipment in Quinault country. The short, stout

"My dad was a bow-maker. He made bows mostly out of Pacific yew. It grew a lot around here [but] for every 100 fir trees, there might have been one yew tree; it wasn't really a prolific tree.... It had little red berries... birds loved the berries. After they reached a certain stage, then they would eat the berries and their droppings would end up at another place for that tree to grow. That's how that tree got to be what it is. There was more so done by the birds than any other way of proliferating. It was slow. There just wasn't a whole lot.... My dad didn't use so much of the red part. The outside was the white part of the wood. It was half and half [red and white]. So that the front part that faced him, that was the strength of the bow and the other white part was the part he would soak it in the oil so that that part would remain pliable. But if he just used the red part, it would break. The backing of the bow was natural. The backing of the bow was made from the white part of the tree. He learned that from his father and of course I learned it from him: how he made his bows."
—Chris Morganroth

sticks, roughly three feet in length, sometimes flatted on one or both sides, helped pry such foods as mud clams and camas bulbs from the ground. Other specialized yew digging sticks were crafted for digging razor clams on sandy beaches.

Yew has been a popular wood for making paddles. Though yew paddles are heavy, they almost never break. Yew is also used in constructing parts of the canoe that must be especially strong, such as the ribs, "rub rail" or gunwale along the top of the canoe, sea fittings, and other parts—some are made from yew saplings that can be form-fitted to the canoe because of their flexibility. Canoe bailers and other essential equipment were also sometimes made from this wood. Yew is used to make pegs that join pieces of wood for fastening canoe parts together, or for building boxes and other goods. To do this, one drills holes in the wood where pieces meet, inserts yew wood pegs and then cuts and sands off the pegs, sealing the hole with pitch-based glue.

Beyond these uses, yew wood has been used to produce durable tools. This wood has been a favorite for crafting tool handles, such as the handles of elk-horn chisels, yielding a handle strong enough to withstand repeated hitting with a mallet. Often, a cord is wrapped around the handle to ensure it does not split. Similarly, yew has been used to make wedges for splitting planks, which were often hit into place with a yew-wood maul, often with an attached yew-limb handle, that is struck with a rock hammer. Most traditional Quinault houses were made with boards split from cedar trees or logs using yew mauls. Yet, yew could also be used for more delicate work. Traditional clothing makers, for example, used a specialized yew tool to press the seams of new cedar bark garments, making them more waterproof and giving them a smart "finished" look.

The richly colored wood is both attractive and nearly unbreakable, making it traditionally popular for household goods as well. Though it takes some time to carve spoons and toothed hair combs from yew, carvers have appreciated how these items can be admired for years to come. One traditional game set consisted of twelve polished yew wood disks, each roughly the size and shape of thick poker chips, one painted black and one painted white. Each opposing side hid their chips in bundles of fluffy cedar bark, trying to determine which of their opponents' bundles contained the marked chips. Even furniture has been crafted from yew wood.

The tough wood can be extremely difficult to work if not cured carefully or steamed. Steaming allows yew wood to be bent into roughly the right configuration before detailed carving. Ideally, harvesters can locate and harvest wood where natural bends in the tree match the general shape of the desired tools, such as digging sticks or bows, as this makes manufacturing easier and lends added natural strength to the tool. Apart from these desirable bends, it is generally important to find wood with clear grain for making large items, as yew wood can often be twisted or gnarled. Gnarled trees are used as well, with small pieces taken to manufacture small items. The tree contains both white and reddish wood—darker wood is the heartwood from the middle of the tree. In fact, the Quinault name recorded for yew, *tla-má-ak* or *k'lam'ma'aq*, literally means "red wood." Bowmakers and other traditional craftspeople choose which part of the tree's wood to gather according to the needs of their project. Both parts are strong, but the white wood is more pliable, for example, and has thus been used for the backing of bows.

A diminutive tree found in the low light of the rainforest understory, yew is the only needled, red-"berried" conifer to be found.

Good wood has always been valuable, with people keeping an eye out for fine trees and sometimes trading out or gifting the wood to carvers. In the nineteenth and early twentieth centuries, items once made from yew, such as digging sticks and other items, began to be forged by metalsmiths, with metal quickly eclipsing the use of yew for crafting traditional tools. Still, the old yew tools were durable. Some families retained old digging sticks and other items for many years. Younger carvers have rediscovered traditional yew-working techniques, bringing yew back to common use.

There are many traditional medicinal uses for yew, most involving bark taken from the trunk and stems. The bark was chewed and placed on wounds. Though this application stings, it is said to accelerate healing. A decoction made by simmering the dried bark was used for lung ailments and for internal injuries or internal "pains." In fact, some sources suggest that Quinault was the only Western Washington tribe to use decoctions of the bark medicinally. A ball of the needles or a handful of small yew twigs were sometimes rubbed against the body as part of ritual and mundane cleansings.

In the late twentieth century, medical researchers confirmed that this tree's bark—containing the compound taxol—has cancer-fighting properties. This launched a brief cottage industry in the Pacific Northwest, with harvesters peeling bark from yew trees wherever they found them. Since many commercial harvesters were not careful to avoid peeling aggressively, many yew trees died in the late twentieth century. By that time, a synthetic version of taxol was available, ending the practice, but the relatively uncommon tree had become even more rare along the Quinault coast.

What It Looks Like

Yew is a small evergreen tree with drooping branches and reddish scaly or papery bark—lanky and spindly in some settings, denser in others. Dark-green needles are soft and flat, and berry-like "cones" are tiny. Yew is the only upright conifer in our area with "berries," which have a thin layer of bright red flesh surrounding a single hard seed like a cup. Male (pollen-bearing) and female cones are borne on different trees.

Where to Find It

Yew is found in older, well-established moist forest, usually near streams at low to middle elevations. Harvesters look for large patches with large trees, so as to avoid overharvest. Woodworkers have their "private gathering areas," not usually advertised widely. Higher-elevation trees are often denser, but also more stunted and twisted.

When to Gather

Yew wood could be gathered anytime, but was traditionally harvested in the mountains in spring and summer. The bark was easiest to peel in spring and summer too, when the sap is running. Because the tree has been so severely affected by industrial wood and medicinal harvests in the past, little to no harvesting is recommended for anyone without a pressing cultural need.

Traditional Management and Care

Yew is very slow growing and has been heavily affected in places by logging and commercial harvesting of bark. Traditionally, understanding this to be an uncommon and slow-growing tree, people harvest trees selectively and respectfully, knowing the old trees will not be replaced anytime soon. Part of why harvesters seek out large patches with large trees is to avoid overharvest that might eliminate particular groves. Bark harvesters should take care to not remove too much bark and avoid damaging the inner bark when peeling loose outer bark pieces. Branches for tools can be cut from living trees, and they will be able to regenerate themselves.

Cautions

Yew trees contain potentially toxic alkaloids. The seeds and needles are poisonous. The bark can be toxic too, if too much is consumed. In the unlikely event that one is seeking to use yew medicinally, consult health-care professionals first and start with very small quantities.

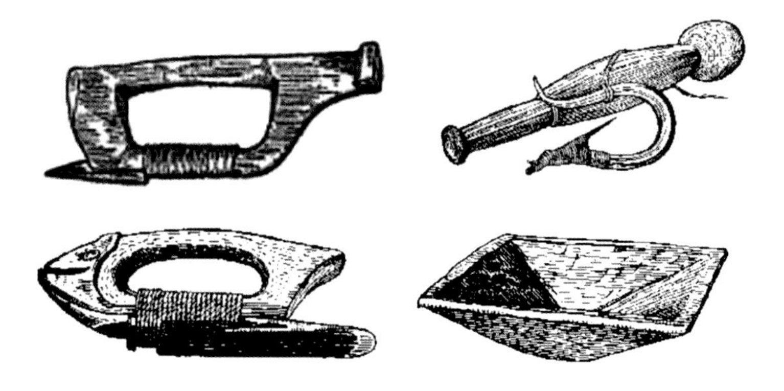

Yew wood is uniquely strong, making it the ideal material for durable traditional tools—from adzes used in woodcarving to halibut clubs and hooks to bailers used in oceangoing canoes.

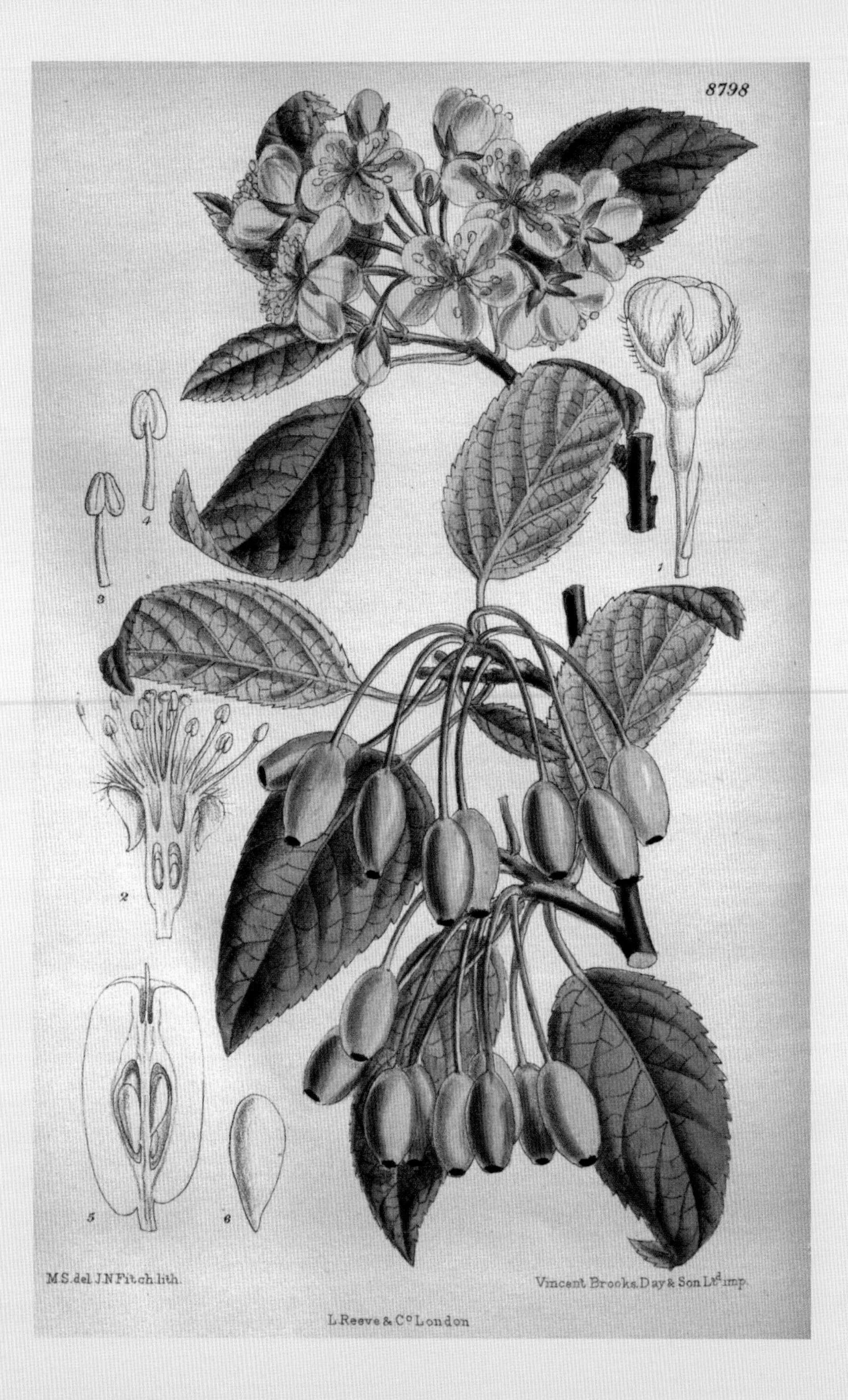
8798
1
2
3
4
5
6
M.S.del J.N.Fitch lith.
Vincent Brooks, Day & Son Ltd imp.
L.Reeve & Co London

Pacific Crabapple

ḳwiʔcaniɬ | *Malus fusca*

Traditional Uses

Wherever the crabapple grows, it is appreciated for its tart fruit, sometimes eaten off the tree, sometimes picked while firm and stored in baskets as the fruit softens and sweetens. Crabapples tend to pick up flavors as they sit, so baskets are best when clean or lined with leaves or plants that impart desired flavors. Beyond this, crabapples are commonly cooked—especially when a bumper crop occurs. Historically, crabapples were wrapped in leaves and pit-cooked, though they could also be simmered with hot stones. Containers of the tart applesauce mixture were then submerged in cold water to preserve the fruit for later use. Today, many families make crabapple jam and jelly, or can crabapples, often mixed with berries to enhance flavor and sweetness. Crabapple mixtures are also simmered on a stove and sweetened like applesauce. Finally, simmered crabapple is used as a natural thickener and gelling agent for pies, jellies, and other foods—such as berry pies that would otherwise be too soft and runny.

In Quinault country, however, crabapple is equally important as a medicinal plant. In particular, the peeled bark is used in many traditional medicines. Tribal members drink teas or strong decoctions made by simmering the bark for kidney and gall bladder disease, and for a range of digestive complaints such as diarrhea, dysentery, intestinal pain, stomach pain, and other soreness within the digestive tract. Crabapple bark decoctions are also popular for reducing menstrual cramp pain, or as a mild antibiotic or "cleanser," taken internally for infections, including bladder infections. In these cases, crabapple bark tea is consumed in quantities befitting the severity of the ailment. Often it is drunk in large quantities for severe or enduring conditions. Teas made from bark and fruits are also used as part of remedies to prevent colds and flu,

"We used to go out and pick a lot of crabapple. That's real easy to pick. I never mess around with it. I just grab a whole cluster of it and chew on it at once. They're kind of bitter, but still it's kind of got a taste that I kind of enjoy chewing on. You have to watch because the bears are always eating them at the same time!"
—Francis Rosander

or to reduce their symptoms. (For example, a crabapple leaf or bark tea has been reported as a remedy for conditions causing a person to spit up blood.) Elders note that this remedy has been successful against persistent and antibiotic-resistant infections like MRSA (methicillin-resistant *Staphylococcus aureus*). Crabapple tea has even been reported to have moderating effects on hangovers. Additionally, Quinault elders report remarkable success in using crabapple decoctions to combat the effects of cancer; a number of individuals report they have seen the lives of cancer survivors extended by the regular consumption of crabapple bark decoctions, often alongside doctor-prescribed medicines.

Crabapple decoctions are used externally to heal infections of the skin, diseases like chickenpox that affect the skin, or wounds. Fresh wounds, especially those that are severe, are sometimes soaked in crabapple bark teas to prevent infection, reduce swelling, and accelerate the healing process. With its antimicrobial properties, crabapple decoctions of varying strengths have also been used as a wash for infected eyes, or as a partial remedy for certain venereal diseases. Boiled crabapple bark, or a compress soaked in hot crabapple bark decoction, is sometimes placed directly on arthritic or overworked joints, helping to reduce pain and swelling.

Families traditionally camp out for a night or two beside especially good crabapple patches to harvest bark or berries—sometimes returning to the same groves year after year, developing special ties to groves that have protected family health year after year. As older trees can have thick and brittle bark, harvesters often focus their bark harvests on younger trees. The harvesting of crabapple, like any plant of great medicinal importance, is traditionally subject to cultural protocols. Like cedar bark, crabapple bark is taken from only a portion of the tree, so as to not kill the plant. People commonly take strips from the trunk that are narrow—roughly three inches wide—or cut off a few nonessential branches from the tree, peeling the bark from those. The bark is rolled up, taken home, and dried. Families often maintain a supply of the bark in

The tart, tiny edible "apples" of crabapple range from green to red, suspended on stiff stems from small trees that have been revered and tended by generations of Quinault harvesters.

their home, drawing from it gradually over time as needed. Teas and decoctions are usually best when simmered but not allowed to boil. Hard boiling not only makes the brew bitter, but may reduce its medicinal value. The brew is usually dark and not especially tasty—"nothing you're going to go drink for enjoyment," as Gerald Ellis put it. Some regular users report that crabapple bark, used once, can often be used again in a batch with fresh water. When it no longer imparts color and smell to the water, it should be replaced.

"They harvest crabapple bark . . . the same [as] with the cedar bark. They tried to just take one peel off, so the tree could heal up. . . . Boy, are they tough [and] we didn't have sharp knives or axes to cut them up. . . . They probably did it trying not to kill the tree: took only one, maybe two [strips] of what they were going to use. The tree would continue growing."—Phil Martin Sr.

What It Looks Like

Crabapple is a shrubby, often spreading tree that can reach up to 40 feet high, but is often shorter, with red-brown to gray scaly bark and sharp spur shoots. Pointed leaves are serrated and finely toothed at the edges, often with a course, pointed tooth on one or both edges of the blade. Clusters of white, five-petaled blossoms become tiny, long-stemmed apples with colors ranging from green and yellow to red and orange when ripe.

"When I was . . . in my early twenties, my grandmother had cancer and my grandfather always made tea for her out of crabapple. My dad and I, we would go out, it was always around April, May, when we noticed that the crabapple was actually blooming, and we'd get some branches and we'd take them. I'd watch my grandfather go and peel the bark off of it, and cut it up into really small pieces—very delicate, he was very precise in what he was doing. I always remember, she was always drinking that tea and she always said that it always made her feel better. And I actually think that it actually prolonged her life. I was in my early twenties at the time and she didn't pass away until I was in my thirties." —Mandy Hudson

With its white, five-petaled blossoms, crabapple becomes especially visible along springtime forest margins, signaling to harvesters each tree's potential as a source of both food and medicine.

Where to Find It

Crabapple can be found in the open forest at low to middle elevations, but it is more abundant in fertile soils along rivers and streams, as well as along the margins of freshwater wetlands, estuaries, and beaches.

"One of my great-aunts taught me how to use that [crabapple] for a pectin, because we didn't have pectin for making jams and jellies. She'd mix it in with her pies—you know, smaller chunks in pies for your pectin that holds it together." —Alicia Figg

When to Gather

Crabapple bark is especially gathered for medicine in mid- to late spring (April, May, early June) when the sap runs and the bark is loosened; by summer, the bark is said to begin to stick to the tree and harvesting becomes very difficult. Fruits are available later in the summer and into the fall.

Traditional Management and Care

Traditionally, Quinault people harvest crabapple bark only selectively, so as not to kill or seriously harm the tree—taking only small strips from the central trunk or peeling the bark from nonessential branches that are removed from the tree. There is some evidence to suggest that burning of meadows allowed Quinault people to increase the number and productivity of crabapple trees on the margins of these clearings, and that Quinault ancestors may have employed other techniques to enhance the output of these culturally important small trees.

"After carving my joints were sore. . . . We went next door. 'Auntie, how do you use this [crabapple] stuff?' And she said, 'Well you take the bark off of there and you let it dry for a couple days and then you boil it. And while it's really, really hot, to where you could barely stand to put your hands in there, you take it, like, on a rag and you soak it up and you wash your arms and your joints and soak your hands in there.' I did that for a couple of days and it really worked. You know my arms, my joints felt really good. And we still do that to this day."
—Guy Capoeman

Cautions

Crabapple has tremendous potential to heal, and may yet provide cures for ailments old and new. Still, crabapple bark users should be very careful with all medicines and ask knowledgeable physicians for guidance, especially when trying to combat life-threatening illnesses such as MRSA and cancer.

The tree has sharp spines—sometimes irregularly placed, hard to see, and tough on the hands of harvesters who grab the tree before looking. Some elders recall epic tales of people who have confused cascara with crabapple when harvesting medicinal bark, often with dramatic and unpleasant gastrointestinal results.

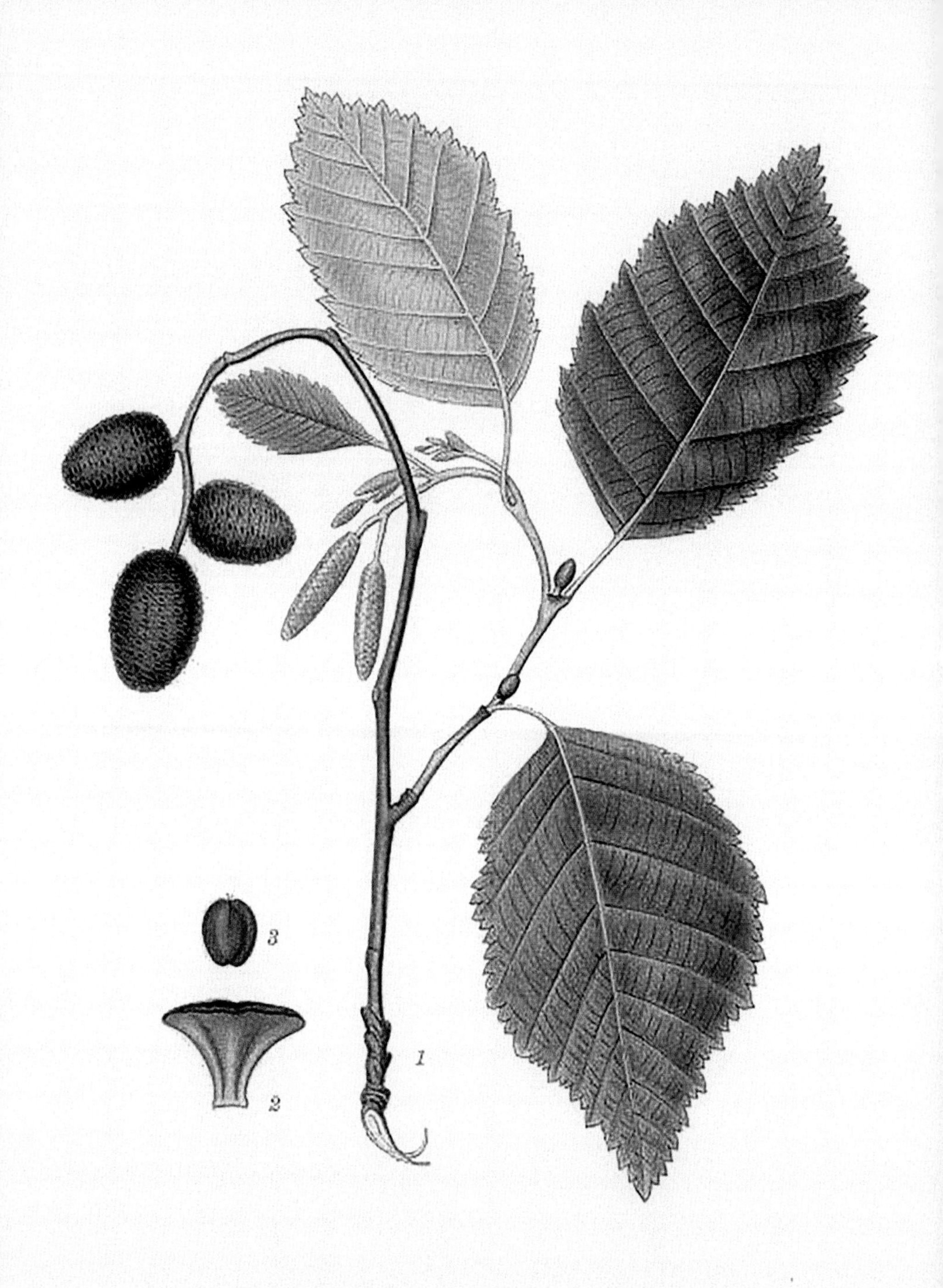

GRÅAL, ALNUS INCANA MOENCH.

Red Alder

malp | *Alnus rubra*

Traditional Uses

Alder is a widespread and multifaceted tree. With root nodules that produce their own nitrogen, this tree can live comfortably in poor soils where other trees cannot. Historically growing along rivers and streams, alders benefited from the advent of industrial logging and farming in Quinault country, which allowed the tree to spread onto disturbed ground, making it a common species in today's forest.

The light, almost off-white wood of alder trees can be used for many purposes. Some say alder is second only to cedar in its importance for carving. With mallet and chisel, Quinault carvers fashion single pieces of alder into canoe bailers, masks, rattles, and many other items. Especially in times past, alder was used to craft eating utensils. Unlike most other woods, it imparted no flavor to the food; thus, a large proportion of Quinault families' bowls, dishes, platters, food boxes, cooking troughs, water buckets, spoons, and other eating utensils were made from the wood of red alder. Although the wood is not especially strong, Quinault people also have used alder poles for the construction of temporary shelters, fish traps, fish-drying racks—even canoe paddles, though these are a bit weak for heavy ocean use. From the hard wood of alder roots, Quinault people crafted fire-starting drills (*dji tstcup*, "twirl fire"), consisting of a dried alder wood shaft above and a grooved alder wood hearth below.

Additionally, alder has always been a popular source of firewood. Relative to other trees with denser wood and more pitch, alder is clean-burning, with less smoke and fewer popping sparks. For this reason, alder is a favorite for smoking fish and other foods. It is also a preferred wood for open fire pits, as in ceremonies or longhouses—settings where people are especially eager to avoid flying sparks and heavy smoke.

White and gray on the outside, the tree's bark is brilliantly reddish-orange on the inside when exposed to the air, giving red alder its name. This bark serves as an important source of dye. Boiled and sometimes soaked with cedar bark, it can impart a brilliant red color to cedar bark strips used in regalia, colored baskets and basket design work, and similar purposes. Other basketry materials, such as grasses, are sometimes dyed in this manner as well. When soaked with alder and hemlock bark, nets made of light-colored plant fibers were stained reddish-brown, making them nearly invisible to fish. At times, animal traps were also dyed this way, including metal traps, taking the shine off the metal and making them less visible. After burning, alder bark sometimes served as the base for black dye or paint that would be mixed with oil or salmon eggs. Alder charcoal, mixed with oil, was used as a paint for the skin and, along with charred flowering dogwood (*Cornus nuttallii*), was a leading source of black pigment for traditional tattoos.

"They used alder bark for your reddish color. Boil it. Let it cool. It gets pretty orange, pretty red—a real pretty color. Like anything else, you can mix one color and another color and get the color you want."
—Justine James Sr.

Beyond its value as dye and paint pigment, alder bark has many other uses. Salmon eggs were traditionally cooked on top of large, concave pieces of alder bark. Alder bark often serves as a liner for baskets used for gathering or storing elderberries—imparting little flavor while also protecting the basket from the nearly permanent discoloration that can be caused by elderberries. So, too, alder bark has been used with skunk cabbage to line pits used to cook elderberries. Finally, the juicy innermost bark and "sap" of the red alder is edible and sweet, and has sometimes been used in lean times, or as a sweetener in traditional desserts such as whipped soapberries.

Additionally, alder bark has been an important medicine—a boiled decoction of its bark used for digestive problems, for colds, for cuts and sores, as a foot bath, and for many other purposes. An ointment made from an alder bark decoction mixed with fat, and more recently beeswax or petroleum jelly, has been used to soothe irritations and sores on the skin. Elders describe alder bark medicines as soothing, antiseptic, and anti-inflammatory, and some keep pieces of bark on hand for this reason. Twigs, leaf buds, cones, and catkins also have many reported

Cooking utensils, canoe bailers, and many other traditional items are made from the wood of red alder.

Young red alder trees, sprouting their deeply veined and serrated leaves, bright green, on thin trunks rising from the forest floor.

Late in the season, red alder leaves still show distinctive red alder serration, while cones—in truth, female catkins—are deep reddish-brown and compact, growing in small clusters.

medicinal uses, and catkins especially are chewed for digestive, respiratory, and skin problems. Quinault people sometimes placed the bodies of the dead in canoes historically, within the branches of large, mature alder trees.

What It Looks Like

Alder is a common deciduous tree reaching up to 120 feet, with white to gray blotchy bark. Shiny oval leaves come to a point and have coarsely serrated edges. Long, hanging reddish pollen-bearing catkins appear in the spring, along with green oval seed cones that gradually turn dark brown and release small, winged seeds.

Where to Find It

Common along rivers, streams, roads, and in regenerating logged sites, red alder is the most visible deciduous tree on the Olympic coast.

When to Gather

Alder wood can be gathered at any time of year, but is lighter and drier at the end of the dry season in late summer and fall. The bark is best gathered in the spring and summer when sap is flowing. Catkins appear in the spring.

A blotchy mosaic of whites and grays mottle the trunks of mature red alder trees; the reddish inner bark is used for medicines and dyes.

Traditional Management and Care

Alder gathered for medicinal or other cultural uses is commonly harvested with an appropriate show of respect. Some people look after the alder from which such materials are gathered and make an effort to help the trees thrive. People who gather bark generally take steps to avoid killing the whole tree, gathering only from limbs or a small portion of the trunk and never "girdling" the tree. At one time, people may have enhanced alder habitat through the use of fire and other means, but modern land disturbance provides ample alder habitat.

Bessa pinx. Gabriel sculp.

1. POPULUS Balsamifera. 2. POPULUS Candicans.

Balsam Poplar. *Heart Leaved Balsam Poplar.*

Black Cottonwood

caʔpac or kalleʔcex̣

Populus balsamifera subsp. *trichocarpa*

Traditional Uses

Lining the banks of rivers and other fresh water, black cottonwoods are the largest deciduous trees in Quinault country. Huge cottonwood logs gather in logjams on rivers like the Quinault River, which men once cleared by hand or fire, opening the way for canoes. Additionally, giant cottonwood logs washed down the rivers and onto the beaches. So big were these drift logs that, in one ancient Quinault story, Raven goes to an underworld where people mistook a cottonwood log for a beached whale, even attempting to butcher and eat it.

Though cedar trees have been the ultimate canoe logs, cottonwood was also used to construct canoes—especially river canoes, hollowed out of single trees. Cottonwood absorbs water quickly, though, so the canoes require sealant. They should not be left in the water for long. In times of war, cottonwood logs were used alongside cedar pickets to build defensive "palisades" around villages. Split down the middle, the logs were set in the ground, round side out, and lashed together with crossbeams of thin hemlock. Cottonwood was also popular for use in fire drills, with a dry dowel spun by hand against a grooved piece of wood below. Cottonwood's durable bark had practical uses as well: large slabs of this bark were sometimes used as roofs and walls for temporary houses, such as at resource harvest camps or for short-term visitors to established villages. The pitchy inner bark and sap is also edible, and some tribes used the soft inner fibers of the bark, twined with other plant materials, to make cords and ropes.

The fragrant sap and pitchy buds of cottonwood have served as an important medicine, used topically as an antiseptic for cuts and other wounds, or in soothing salves. Sap from cottonwood burls, or extracted from bark gathered at ground level, is sometimes said to be

"Before the introduction of matches fire was procured by friction from very dry dead cotton wood. A stick of this was pointed and placed in a small cavity made in another piece of wood, the hands rapidly moving the upright stick as if drilling. The sticks with three cavities were placed upon the ground, the Indian kneeling and placing a knee upon each end. He placed one end of the smaller stick in one of the cavities, and, holding the other end between the palms of his hands, kept up a rapid half-rotary motion, causing an amount of friction sufficient to produce fire. With this he lighted the end of the braided slow-match of cedar bark. This was often carried for weeks thus ignited and held carefully beneath the blanket to protect it from wind and rain."—Charles Clark Willoughby, 1886

especially potent. An infusion made of cottonwood sap, or by boiling the bark gathered on the tree near ground level, has been consumed for respiratory ailments including tuberculosis. This decoction can also be used externally for irritated skin, eyes, and throats, to soften the hair, and for many other purposes. A variety of other medicinal uses have been described; for example, some sources describe boiling the bark to make a gargle for sore throats, while others describe burning the buds and bark to charcoal and using it in stimulants and expectorants. Poultices are sometimes made from crushed cottonwood leaves.

What It Looks Like

Black cottonwood is a deciduous tree reaching 100 feet, with an elongated or triangular profile and shiny, smooth-edged, long-stalked oval or triangular leaves that come to a point. Young bark is smooth; mature bark is deeply furrowed and gray. Distinctive leaf buds in the springtime are orange-brown, sweet-scented, and glisten with nearly dried resin.

Oval to triangular black cottonwood leaves emerge from sweet-smelling resinous buds, as spindly limbs spread outward from a sturdy gray trunk.

Special shallow-draft canoes, designed for traveling the coast's small log-choked rivers and streams, are traditionally carved from black cottonwood trees.

Where to Find It

Cottonwood is found in moist forested sites, especially along the major rivers of the Olympic coast, but also along streams, lakes, or in swampy depressions. Occasionally found in isolated drier sites.

When to Gather

The wood of black cottonwood can be gathered anytime. Buds and liquid sap are best gathered in the spring and summer, while hardened sap can be obtained from the tree at other times of the year.

Traditional Management and Care

Black cottonwood gathered for medicinal use or for important purposes, such as canoe-building, is traditionally gathered with an appropriate show of respect. Some people look after the cottonwood trees and groves from which such materials are gathered, taking steps to help the trees thrive.

Cautions

Some tribes have oral traditions suggesting that medicinal concoctions from cottonwood should be taken only in moderation and can be toxic in high doses.

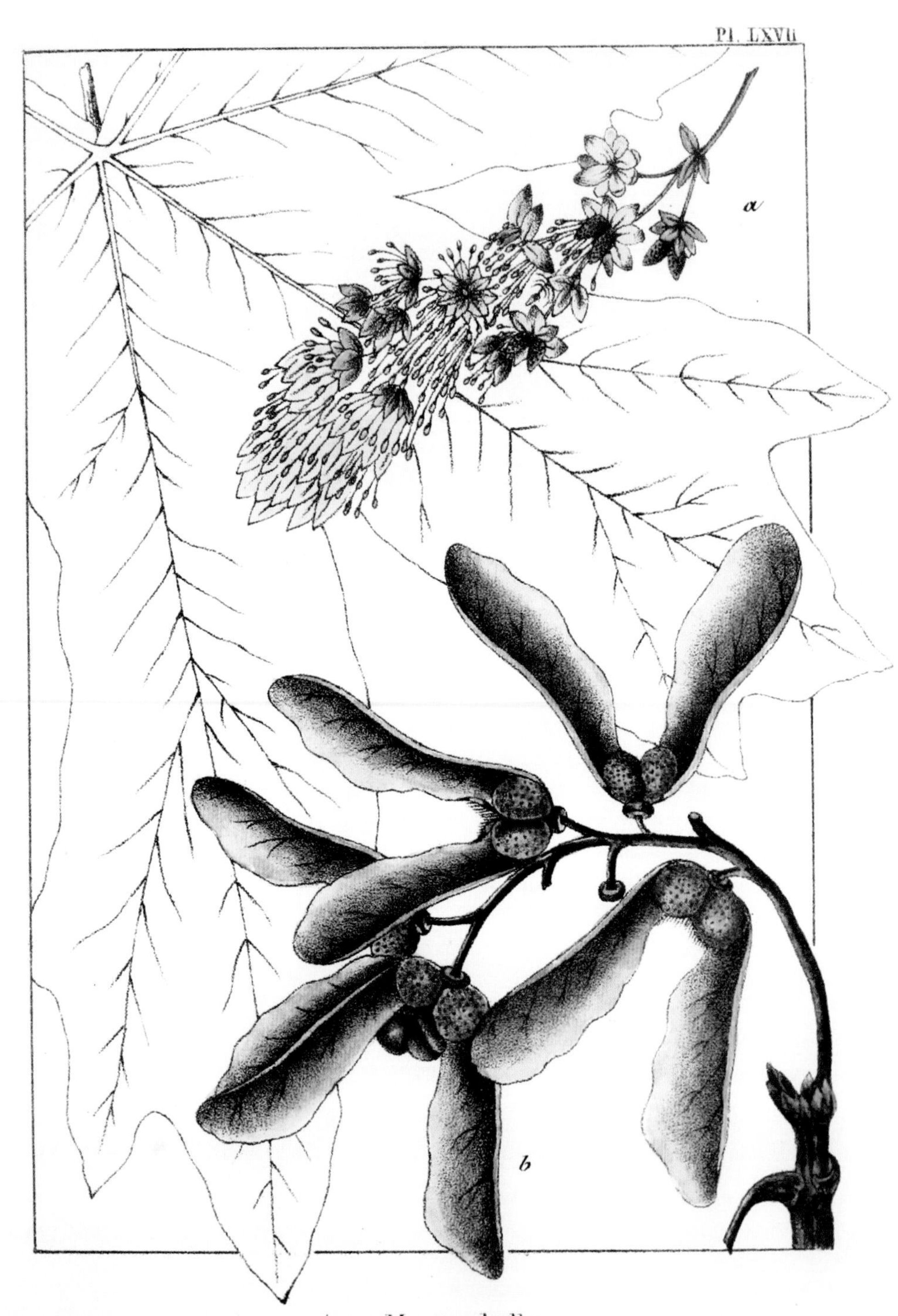

Acer Macrophyllum.

Large leaved Maple. *Erable à grandes Feuilles*

Bigleaf Maple

naʔustapp maɬp | *Acer macrophyllum*

Traditional Uses

Bigleaf maples, with their moss-strewn branches and giant green leaves, tower above the riverbanks, streams, and other damp inland places of Quinault country. The hard, dense, clear-grained wood of this tree is widely popular for manufacturing traditional goods. Bigleaf maple is one of the most popular sources for carved paddles—its strong but lightweight wood is a perfect match for the demands of hard ocean paddling. Such properties make maple a popular wood for any item that needs to be tough: spear heads and fish clubs; cedar-bark-pounding tools and "hackler" bark shredders; canoe bailers and utility buckets; and tools such as bobbins and seine blocks for making nets and heavy mats. Carving usually requires a mallet, a chisel, and patience. Beyond its strength, bigleaf maple wood has an advantage over many other woods in that it does not impart unpleasant flavors or odors to food. For this reason, maple is also carved into bowls, dishes, platters, spoons, ladles—sometimes even huge feast platters, carved like small dugout canoes from maple logs up to roughly ten feet in length.

Maple wood has many other values. Quinault historically made black paint by mixing whale oil with bigleaf maple wood charcoal. A decoction of the bark has been used by some tribes for tuberculosis and other lung ailments. Maple wood is also popular for smoking fish. The burning wood has an especially pleasant smell and, lacking heavy pitch, seldom pops or throws sparks. Traditionally, people gathered downed maple wood from riversides and beaches where the wood had dried and cured perfectly for smoking. Some have said the flavor of maple-smoked salmon is the preferred traditional salmon flavor of Quinault people.

Other parts of the tree had value in food preparation as well. The leaves are sometimes used in cooking, especially to wrap elderberries or camas bulbs for pit-cooking and storage. Large numbers of leaves are sometimes spread on the ground to make temporary "mats" to keep fish clean during processing. Parts of the tree are edible: the soft inner bark and "cambium" was traditionally eaten fresh or dried, while the sap was also eaten and, occasionally, boiled down to maple syrup.

"We went up to Maple Camp. That's why it has its name I guess. And we took a jet boat up there. . . . We felled a maple tree, a big one. . . . We dropped it in there and we milled it into planks. . . . We cut it into sections and we hauled them out and we made twenty-four paddles out of that. . . . Those make really good paddles as well: strong, flexible, dense." —Guy Capoeman

Over time, maple trees have become less common on parts of the Olympic Peninsula. Elk and deer browsing on saplings have increased, due in part to the extirpation of wolves and other predators. Maple is also a popular wood for making guitars and other musical instruments. In recent years, the Quinault Indian Reservation has been the target of poachers from off-reservation who illegally cut down large trees to sell to musical instrument manufacturers, prompting expanded policing across the reservation.

What It Looks Like

Bigleaf maple is a large deciduous tree, growing 50–150 feet tall. Classic large green maple leaves, long-stemmed with roughly mitt-shaped blades, turn yellow and spotted in fall. The tree crown has a rounded shape, gray slightly furrowed bark, and produces dense clusters of pretty dangling yellow-green flowers that turn into dual-winged seeds. In Quinault country more than in other parts of its range, bigleaf maple branches and trunks tend to be draped in moss and to host patches of licorice fern.

Where to Find It

Bigleaf maple is found along the major rivers, as well as along smaller streams. Where the soil is damp but well-drained, it also can grow in denser forests away from flowing water, at low to middle elevations.

When to Gather

Bigleaf maple wood and limbs can be gathered anytime. Downed wood is generally used when its moisture content is right for the intended use. Sap and bark are most commonly gathered in the spring. Leaves are used when available, from mid-spring through fall.

Traditional Management and Care

With bigleaf maples, as with other trees, harvesters have sought to not kill trees unnecessarily for wood, which is among the reasons people rely heavily on fallen deadwood. Gathering of sap, bark, and leaves is usually done respectfully, taking enough to not harm or kill the tree. Poaching and reduced seedling survival remain a concern both on- and off-reservation.

The deeply lobed leaves of the bigleaf maple—the largest of coastal trees—are distinctive and are traditionally used to create temporary mats of food containers.

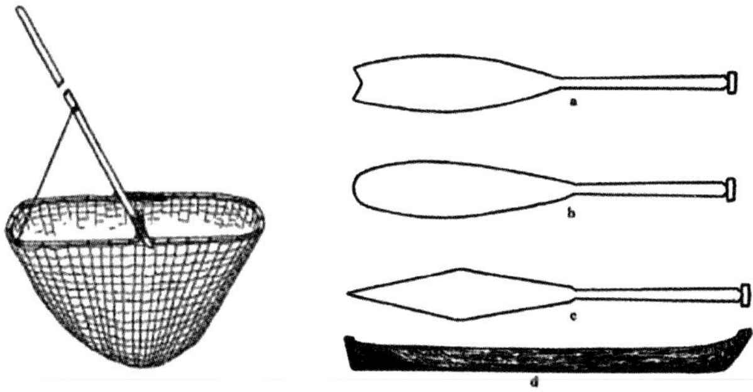

The durable wood of bigleaf maple has been used to construct many important tools—paddles, dipnet handles, even the occasional canoe.

Vine Maple

maxa?attcalnix or malp niɫ | *Acer circinatum*

Traditional Uses

Winding through the understory of deep coastal forests and crouching in dense thickets along rivers and streams, vine maple is well known to anyone who travels in Quinault country. Vine maple is also a culturally important tree—so important to Quinault basketry traditions that one term for the tree is *maxa 'atca nix,* literally "basket tree." A village fronting the Quinault River about six miles upstream from Taholah was named La lcil, "Vine Maple Place," reflecting both the historical abundance and importance of the maples along this part of the river.

The vine maple tree's long, straight shoots are very flexible and strong. Cut near their base, the shoots are traditionally peeled of their bark, then scraped and split to make baskets and many other durable goods. Woven together, vine maple shoots are used to construct large burden baskets for carrying heavy loads like firewood gathered on the beach or freshly caught items like clams, roots, or fish that needed to be cleaned. Vine maple baskets often employ an openwork basket style—one common type was about two feet long, with a cross-warped twine or checkerboard weave. Historically, these large baskets were often worn on the back with a tumpline around the forehead. A smaller basket of similar design, with smaller gaps, was made for carrying berries, a style of basket often "twilled" with a diagonal weave. With this coarse open weave, the baskets could be submerged in water, allowing harvesters to clean their clams, berries, or other goods, allowing dirt, twigs, leaves, and other unwanted debris to wash away. These were utilitarian baskets, not often decorated.

Vine maple is also used to build fish weirs (*skeli h*) to catch salmon and trout along the Quinault and other rivers. Maple has been especially important as the structural "brace poles" of some fish traps, the wood lasting a little longer in wet, turbulent shoreline conditions than other woods. Long straight trunk sections, four to twelve feet long, are stripped of branches and sharpened at one end to secure them in the mud. Within these outer structures, weirs are woven like long, conical baskets, with cedar branches woven between smaller wattle sticks of hemlock and vine maple—roughly one-half to three-quarters of an inch thick—that were stuck into the river bottom below. Fishermen sometimes built fishing platforms near these weirs of vine maple and hemlock poles, so they could use dipnets to push fish toward the traps. When fish runs were known to be imminent, fishermen harvested new vine maple poles for these purposes.

"They made the big baskets from vine maple then wove them coarsely and made bigger baskets for carrying. They used vine maple and spruce, cedar roots to weave them. They even used them to cook in."
—Horton Capoeman

The strong, springy wood was used for many other purposes as well. A traditional Quinault style of drum is made of smooth deer hide, stretched over a circular loop of vine maple. These drums are light and portable, taken for drumming on canoe journeys and used for other purposes—especially when drumming away from home. Snowshoes for trekking into the mountains were made with ovals of vine maple, tied together with lashings and attached to the feet with thongs of elk skin. Relatively long, thick vine maple poles were sometimes used to hold cedar plank roof boards in place on longhouse roofs; these ran parallel to the rafters and were often tied to roof planks with cedar withes. Carvers have occasionally fashioned spoons and other utensils from vine maple, though bigleaf maple wood was preferred for these uses. Vine maple is occasionally used for the construction of bows. Not as strong as some other woods, the maple stave is often reinforced with sinew or other strong but springy materials. Sometimes vine maple has been used to make shinny sticks. People looked for sticks with a natural curve at the end, roughly as long as from the middle of the chest to the tip of the middle finger. Sometimes, a scrubber made of braided vine maple was used to scrub one's body as part of ritual cleansing. Finally, a springy branch of vine maple was used inside longhouses to suspend babies' cradles, giving them a gentle bounce, with all four cradle corners suspended by a cord from the end of the pole.

Vine maple was also a standard source of black paint—the wood was burned and the charcoal mixed with oil or other substances. If vine maple pigment was to be used to paint wood, the charcoal was mixed with salmon eggs or liquid extracted from cedar bark as fixatives.

Strong, springy woods such as vine maple, willow, and spruce root are foundational to woven goods—such as tumplines and burden baskets, used to carry heavy loads of firewood and food back to villages and camps.

What It Looks Like

Vine maple is a small to medium-sized tree that sprawls in the shade and clusters in the sun, with springy branches and smooth bark. Delicate bright green maple leaves, with seven to nine pointed lobes, are slightly hairy, turning gold to red in autumn.

Where to Find It

Vine maple thrives under heavy forest canopy, but is also found in openings along streams or on recently disturbed sites at low to middle elevations.

When to Gather

Shoots for baskets are often cut in the summer and early fall, when the year's new growth is fully developed but still springy and flexible.

Traditional Management and Care

People traditionally seek long, straight shoots of vine maple, but untended vine maple usually gets curved and gnarly. For this reason, people traditionally prune or use other techniques to cut back maple, causing it to grow back long straight shoots in the year ahead. Fire may have also been used to enhance vine maple productivity. When cutting shoots, harvesters are usually selective and are careful to avoid taking too many and harming or killing the tree.

Smaller and more graceful than other forest maples, vine maples have leaves with shallow, pointed lobes and small, paired seed-bearing "helicopter" fruits of bright green to red.

Willows

laleah-kileč

Salix hookeriana and other *Salix* species

Traditional Uses

Many varieties of willow grow in Quinault country, especially along rivers, streams, bays, and beaches. Some, like the abundant Hooker's willow along the coastline, have been a key source of weaving material, wood, and medicine. The long straight shoots that appear each year are harvested by weavers. Bent and secured in suitable shapes, these serve as the foundational "skeletons" of certain baskets, baby boards, and even hats made of cedar bark or spruce roots. To produce fish traps, fishermen often stuck these withes into bay- and river-bottom mud, weaving them together or interlacing them with other materials, resulting in a basketlike trap from which fish could not easily escape. Tied to a bone barb, carved pieces of willow have served as lures or "plugs" for marine fish like rockfish, halibut, sole, and flounder; because of willow's persistently bright whitish wood, fish could see the willow lures even in murky water. Shafts of willow were sometimes used as "fire drills" as well, rotated quickly against larger pieces of wood, such as cottonwood, to ignite cedar bark tinder.

The bark of willow also has many uses. Long strips can be peeled and used as a coarse or temporary string, either in individual strands or with several strands twisted to the desired strength and thickness. (Willow root was also occasionally used as temporary string.) Even today, willow string is sometimes used to tie roasting sticks together when cooking salmon over open fires. When several strands are wound together, willow string can become impressively strong. These heavy strings were wound to make tumpline straps, allowing people to carry heavy loads on their backs with the straps secured across their foreheads. Willow string was also used to make rock-throwing slings, especially when temporary slings were needed on the go, or for use as toys. Heavy

"Willow bark—that was used like string, too. That's what was used on fish sticks [for cooking salmon]. To keep them tied together before we started using wire. ... We did it a couple of months ago. We didn't have any wires, so we grabbed some strips of willow and tied our sticks down. It was right down on the ocean, where we were cooking our fish. [Willows] flourish pretty nice out here, too. I believe they're the first sign of spring plant life."—Micah Masten

willow twine was even braided together to make traditional Quinault swing sets. People took turns sitting and swinging on a braided willow rope hanging beneath a raised crossbar log. Winding these strands thicker still, Quinault hunters could make twine so strong it was used for sea lion harpoon cords, with twenty-foot lengths of willow twine attaching a harpoon point to a long shaft of Douglas-fir.

In addition, willow bark has been an important source of medicine. Throughout the world, people recognize that the bark has potent anti-inflammatory properties. This relates to the presence of salicin, a chemical compound similar to aspirin and named after willow, or *Salix*. In fact, willow bark was once used as a source of commercial aspirin. The ancestral Quinault were well versed in willow's medicinal use. Simmering the bark in water, people have produced a tea that serves as a general health tonic and as a painkiller for headaches, backaches, arthritis, and many other complaints. Some also made a decoction of willow bark for upper respiratory problems such as sore throat, for lung complaints like tuberculosis, and for skin conditions. A light decoction was sometimes used as an eyewash. Simmered with horsetail, willow bark decoctions were consumed by women to regulate menstrual periods. Making the most of its anti-inflammatory value, the bark was also mashed in water (commonly, ocean water) to a mushy consistency and applied to cuts to close the wounds and speed healing. Many other willow bark remedies have been reported, some kept by individual families over generations.

What They Look Like

Willows are diverse, but all have springy younger shoots, ellipse-shaped to oblong green deciduous leaves, and fuzzy catkins making the "pussy willows." Hooker's willow, the most common willow in many parts of the Olympic coast, is a shrub reaching up to 18 feet with stout, springy twigs and long, oval-shaped leaves that are hairy when young. The catkins—both pollen and seed-producing—emerge in spring on separate trees.

Where to Find Them

Willows grow mostly at lower elevations, along the edges of wet areas such as rivers, ditches, and marshes, as well as in moist forests or places with fresh subsurface water along beaches and dunes.

When to Gather

Willow withes can be gathered anytime but are often harvested in the late spring and early summer when they are strong but still flexible. For purposes that require stronger withes, such as for the main hoops on some baby boards, people harvest the prior year's shoots.

Traditional Management and Care

Harvesters sometimes have pruned willow so that it grows long, straight shoots in the year ahead. Fires, intentionally set to manage plant communities, likely contributed to this effect as well.

Clusters of small, bright green leaves erupt from the side of flexible willow stalks in the springtime.

Upright catkins, the flower of the willow, appear on stalks in spring, some species' catkins becoming so fuzzy they bear the common name "pussy willows."

1
2
3
A
B

Cascara, or *Chittem*

x̣wix̣wiʔnil

Rhamnus purshiana
(Frangula purshiana)

Traditional Uses

Cascara bark is one of the most effective natural laxatives to be found, and has long been gathered for both traditional medicinal use and for commercial medicines. Cascara laxatives were used not only to treat constipation, but also to rid the body of parasites or to settle certain kinds of indigestion—the dose being adjusted for the nature of the complaint, the age and weight of the individual, and other factors. Traditionally, people either chewed the bark or simmered it and drank the infusion. Smaller trees and larger trees were said to have different properties, the larger and older trees often having more powerful laxative properties. Furthermore, the wood was occasionally used for tool handles and other purposes. The term *chittem,* widely used in the Northwest for this plant, is derived from the language of the Chinook, who first introduced the plant to many travelers.

In recent generations, a number of tribal members have secured employment peeling the bark for commercial buyers who provide cascara to the pharmaceutical industry. At times this has been an important supplementary seasonal source of income. Fresh bark pieces were sold for a set price, while dried bark was usually worth more per pound; some Quinault families have had drying racks in their yards for this purpose.

What It Looks Like

Cascara is a straight deciduous tree reaching up to 30 feet, with oblong deeply veined leaves up to 6 inches long, similar to alder leaves but with smooth edges. Loose clusters of small green-white flowers turn to round black berries, enjoyed by bears and pigeons.

"I remember one time, one summer, we were all broke. Nobody had any money. Big dance going on at the Wigwam, and Bill's wife and mine wanted to go to the dance. 'Ah, let's hop in the canoe and go down and peel bark.' So we're down there peeling bark. All one day, we peeled all one day and I and Bill, between the two of us, had 300 pounds I think we was getting seven cents a pound. Freddy Cole had 1,500-pound bunch. That same day, he was peeling right alongside of us. He was just that much better than us. He did it every year. He did it every summer. Peeled bark all over."—Gerald Ellis

Where to Find It

This tree is found scattered widely through the Olympic coast in moist, shady areas such as mixed forests or the edges of clearings at low to middle elevations.

When to Gather

Cascara bark is peeled mostly in the spring and summer, with mid- to late summer being peak times.

Traditional Management and Care

Harvesters traditionally make an effort to not peel so much bark from a single tree that it might kill the tree. Burning of clearings may have enhanced the output of cascara on the forest edge historically.

Cautions

Use caution when handling cascara, for even handling the bark can cause a strong laxative effect. Even drinking from streams lined with cascara is said to have interesting results. Anecdotes suggest that it is not a good idea to roast food on cascara sticks. The berries are not poisonous but are reported to have the same laxative effect as the bark.

The leaves of a young cascara tree, oval and bright green, emerge from the forest understory where its seeds have been deposited by band-tailed pigeons and other birds.

Dark red to purple to black, cascara berries have occasionally been eaten by Quinault people, but are mostly left behind for bears and birds.

Deeply veined oval leaves and gray-barked trunks mark cascara stands that endure, even after generations of respectfully selective bark harvests.

Shrubs for Food and Medicine

N° 534.

Ledum latifolium.

W. Miller del^t. G.C. sc

Indian Tea

nəwak̓wa'nti or sɬəgwəlmiš ti

Rhododendron groenlandicum (formerly *Ledum groenlandicum*)

Traditional Uses

Known by common names like Indian tea, swamp tea, and others, this shrub is among the most widely used medicinal plants in the Quinault world. Many keep pots of Indian tea simmering in their homes year-round, or brew a pot whenever someone in the house becomes ill.

In each case, leaves are simmered in a pot until the color and strength of the brew reaches desired levels—lighter for everyday use as a beverage and darker for medicinal use. When brewed and consumed as a richly scented everyday tea, this plant is said to promote general healthfulness and a calm, clear state of mind—consumption is common practice in many Quinault households. When used in larger quantities and at greater strengths, the plant is said to cure ailments. For example, decoctions are widely reported as a remedy for inflammatory conditions; thus, Quinault elders commonly consume Indian tea for arthritis and other joint ailments, as well as for a range of cardiovascular conditions with inflammatory dimensions. Consumed in large quantities, the tea is also said to be deeply cleansing, a diuretic that purges the body of toxins and purifies the blood and kidneys. Reflecting both its cleansing and anti-inflammatory properties, Indian tea is consumed by women after childbirth to help rebuild their strength.

Long leathery leaves of Indian tea extend in all directions from a central stalk.

Elders report that, when consumed regularly, the tea also helps protect against colds. Some families drink extra Indian tea as cold season approaches; and when consumed after contracting a cold, the tea is said to reduce the cold's length and severity. The vapors are likewise healthful at these times. Especially when congested by colds, flus, or allergies, people can breathe the vapors directly, with a towel draped over their head and the pot to capture the fumes. This often works as an expectorant,

"You just take [a little from each] part of the bush there and each limb. That's about all... just don't take them all off. Leave some on the bush. Take all the original leaves off it's going to die. You wipe it out and come back next year it won't be there.... Just look for the orange underneath [the leaf]. Should be a little stronger flavor when you start having the real orange underneath it. That's the ones you try to get. The other [leaves] still have flavor but you use a lot more." —Jared Eison

reducing inflammation and allowing mucous to flow freely, clearing the head and chest.

Starting in the spring, elders gather large bags of Indian tea leaves and sometimes leafy twigs, often continuing to gather through the summer. Families can gather large quantities in a single trip—enough to last them for everyday teas and medicinal preparations throughout the entire year. Harvesters especially look for newer leaves that have the characteristic "hairs" on the underside of the leaf—orange in older leaves, white in younger. Quinault tradition demands that harvesters take only a few such leaves from each branch, and not too many from any one bush, to avoid injuring or killing the plant. Once gathered, the leaves must be well dried, lest they mold. Harvesters spread out the leaves to dry indoors, or outdoors if the weather is hot and dry; traditionally, some even place the leaves near fire to speed the process. Today, many harvesters spread leaves on a tray in a low-temperature oven or dehydrator.

Historically, Quinault people burned over wet prairies and the edges of wetlands to increase the natural availability of Indian tea. This created large patches of the plant—so clearly "cultivated" by the ancestors that some elders traditionally refer to these prairies as the "tea fields." As Chris Morganroth summarizes, "It was the people, it was the human intervention mainly that kept those tea fields wide open." The plant is so closely associated with fire-managed prairies that the Quinault name most commonly recorded for Indian tea, *nūwaqwa ntī,* literally means "prairie tea," and modern Quinault harvesters often use the English name "prairie tea" for this species today. The Quinault Indian Nation now seeks to bring back traditional fire management to prairies within the reservation, in part to maintain the size and productivity of these culturally important Indian tea fields.

What It Looks Like

Indian tea is a low evergreen shrub covered with narrow, oblong leathery leaves with hairs on the underside—white in younger leaves, rusty-orange in older leaves. Clusters of five-petaled white flowers sit, umbrella-like, atop thin woody stems that range from orange-brown to green in color.

Picking Indian tea on the margins of a wetland prairie, traditionally burned and tended to enhance the growth of this important plant.

"When you go out to gather—whether it's Indian tea or going out and getting sweet-grass to help make the baskets—you just follow through and do the work, and do it in a happy mode. You don't have to be Mary Poppins. You don't have to sing or do anything like that. You just have to be in a good frame of mind. . . . And there's another plant that looks just like [Indian tea, so] you have to make sure that you're getting the right stuff. And then make sure you keep it dry. My grand-mother always told me that's the best tea to have when you're actually sick. She used to have a pot of tea on her fireplace all the time when we'd go there for lunch. You could just smell it when you're walking in the door."—Mandy Hudson

Where to Find It

Regionally, Indian tea can be found in bogs or forested wetlands, on acidic, nutrient-depleted soils from low to middle elevations, along with bog cranberries and sphagnum, or peat mosses. Though most wetlands are not actively burned today, remnant patches of Indian tea are often associated with lands formerly managed in this way by Quinault harvesters.

When to Gather

Many families especially gather Indian tea in the spring-time, when the new leaves appear, but other harvesters report gathering later, well into late summer.

Traditional Management and Care

Indian tea patches are traditionally burned and cared for in other ways, allowing natural small patches to expand to become large "tea fields" on open wet prairies. Also key to the success of Indian tea patches is the practice of respectful selective harvesting, with harvesters taking only a few leaves from each branch, and never too many

"What you do [when picking Indian tea] is just bag it up. And what the seniors used to do is get a big pot and put it on the stove; keep on adding water to it and it just gives the whole house a different aroma. Really good for the starting of an ailment, flu and stuff like that: kept it down. Pretty mild for the people in those ways. Not a fruity aroma, not a hard aroma—kind of mellow. Not spicy, but something in between, there. Nice aroma, like a rich candle."
—Conrad Williams

from any one bush. Thus they avoid killing or compromising the plant.

Cautions

Many rhododendrons are toxic, because of the presence of compounds called diterpenoids—especially the flowers, and even honey derived from the flowers. When consumed in moderation, these are usually not harmful. Flowers should be used with caution, if at all; though it is relatively rare to have a reaction to the leaves, anyone experiencing discomfort drinking normal Indian tea may want to reduce or discontinue use.

Clusters of showy white flowers remind us of Indian tea's distant kinship with ornamental rhododendrons.

Dark leathery green on one side, the base of each leaf is densely "hairy," its colors orange to white.

In large, traditionally managed prairies, Quinault families collect and air-dry large quantities of leaves each year for brewed beverages and for medicinal use.

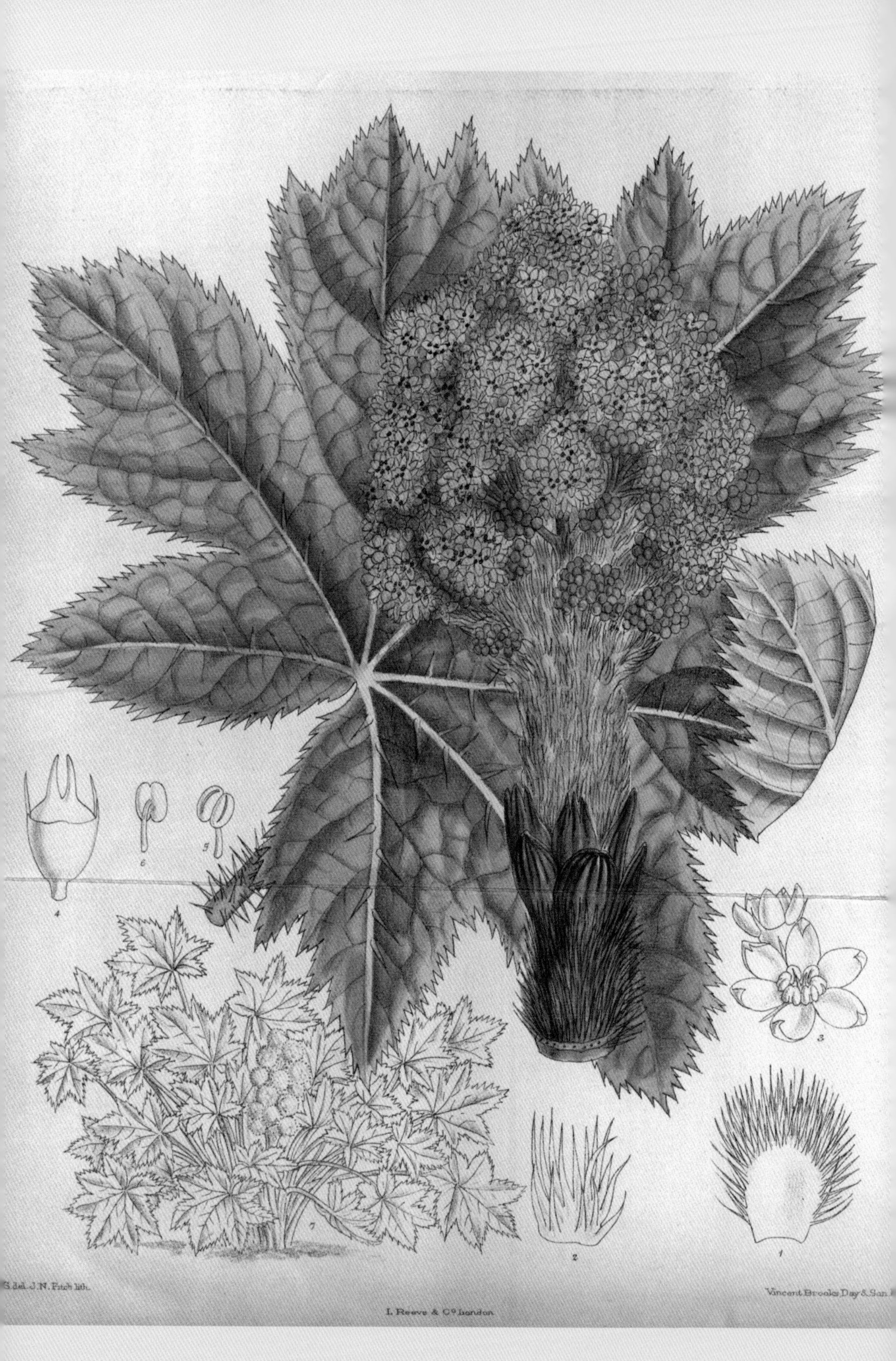
J. N. Fitch lith.
Vincent Brooks Day & Son
L. Reeve & Co. London

Devil's Club

sǩaəkač | *Oplopanax horridus*

Traditional Uses

Among the many plants used for medicinal and cultural purposes by Quinault and other tribes, devil's club is among the most important. This potent relative of the ginseng has myriad uses in Quinault tradition—some having endured uninterrupted for centuries and some that are now being rediscovered.

Devil's club bark tea is consumed to strengthen and energize, physically and otherwise, especially when feeling sluggish, drained, or vulnerable. The plant has been used for protection of the mind, body, and spirit. Teas or decoctions made from devil's club are widely used to soothe inflammatory conditions such as arthritis, for colds, or as a blood thinner and "purifier." They have also been used as a diuretic, consumed to cleanse the urinary tract. Likewise, elders such as Justine James Sr. recalled that a traditional Quinault medicine made of pulverized devil's club bark from the root, steeped in hot water, worked even better than conventional medicines to restore his health after a major heart attack—extending his life and vigor for another four decades. Elders and scientists increasingly agree: if used correctly, devil's club may even help offset the symptoms of modern sicknesses such as diabetes and leukemia. Recipes for traditional medicinal decoctions and teas mixing devil's club with other medicinal plants are remembered today. For example, for pain, devil's club can be mixed with Indian tea, then steeped and consumed as a tonic. Mixed with cascara bark and prince's pine (*Chimaphila umbellata*), devil's club decoctions have been used for such ailments as tuberculosis, or to reestablish regular menstruation after childbirth.

By name and reputation, the medicinal devil's club is widely known for its imposing dense spines.

Topical uses of devil's club are also widely popular and effective, especially drawing on the plant's anti-inflammatory properties. Soaking in a devil's club

"They mix it into a salve and put it on arthritis. I tried it once because my hand swelled up when I went digging. I used it on my hands and went out digging and came back and I could remove my ring. So it helped out pretty good. . . . I shared it with my aunt and she had something wrong with her hand and she put it on there and . . . it worked on her hand, too. For the skin."
—Dennis Martin

decoction, or mixing it in a bath, is popular for shingles, arthritis, and other inflammatory conditions. Salves made from such decoctions are rubbed directly onto arthritic joints, or on skin and joints suffering from other types of inflammation. In recent times, some tribal members have produced salves and tinctures for these purposes—to serve as family medicines, as gifts, or occasionally to sell. The powdered inner bark from the roots of devil's club, and possibly the steamed leaves (with prickles removed), are used in poultices and salves for bruises and eczema. Poultices, placed on the breast, may have been used to regulate mothers' milk flow. Devil's club bark has been reported as a deodorant, possibly reflecting some antiseptic value. The berries, pulverized and rubbed on the skin, have been used as a traditional Quinault insect repellant.

For medicinal uses of devil's club, the green inner bark of the plant is especially used. Some harvesters gather this inner bark from the plant's spiny stalks, while others gather the roots. The brown to beige outer bark and spines are scraped off, and the green inner bark is sometimes peeled off, and even pulverized, before use. Some harvesters simply peel off the spines and outer bark, but do not bother removing the inner bark. Instead they place whole root or stalk sections in a pot to simmer. Most commonly, elders simmer the freshly harvested inner bark or pieces of the root or stalk where this bark is exposed, for an extended period—commonly between twenty-four to forty-eight hours—at a temperature just below boiling. This produces a rich, strong-smelling decoction that can be drunk, mixed into fat or oils to make a salve, or used in other ways. Stalks or roots can also be saved for months, being scraped and boiled only as more medicine is needed. In recent times, some tribal members cut very small pieces of bark and freeze it in plastic bags for later use.

"Devil's club is used for strengthening your body, your energy. We'd take the outer shavings of the stalk and we'd boil that and drink it. And then we put the stalk, the dried stalks away for a later use, for the same thing."
—Guy Capoeman

Alongside all of these important medicinal uses, devil's club has everyday value as well. Elk, and sometimes people, eat the tender and spicy new shoots that grow from the top of the plant each spring. They can be eaten fresh or cooked to taste. The peeled inner wood of thick devil's club stalks has been used for walking canes, as well as being carved into fish shapes, attached to barbs,

Students of Taholah Elementary and Middle School, harvesting devil's club on a class trip.

and formed into lures for catching coho salmon, rockfish, and other species. A small number of elders report eating the red devil's club berries, though most people consider these inedible. Oral tradition suggests that bull elk were considered to be optimally fat when the devil's club berries turned red, signaling the time for late summer hunts. Fishermen sometimes rubbed the entryways to their fish nets and traps with devil's club leaves and carried amulets carved of the wood.

What It Looks Like

Devil's club is a loose, spreading shrub with woody, extremely spiny stems and large maple-like leaves. Cones of white flowers give way to cone-like clusters of small, inedible red berries.

Where to Find It

Devil's club is especially found in moist, dense woods near streams, wetland margins, or on well-watered hillsides at low to middle elevations. Some harvesters walk river or stream banks, gathering roots exposed there. The mountains remain a popular place to gather.

When to Gather

The properties of devil's club are said to change slightly with the seasons. Early spring, and fall to winter, are popular times to harvest roots, whereas spring and summer are often the time to harvest stalks. The fresh edible shoots appear in the spring. In light of their slow rate of growth, any plant or patch of plants should be harvested only sparingly from year to year, allowing them to recover.

"Use the root. Reach under and cut the root off and then you take and dry it and take all the sap out of it, all the bark, and pound it down to powder. I had a real bad heart attack in 1978. They gave me that nitroglycerin. I didn't like that. My face was kind of twisted. I went over and talked to one of the old Indians and tell him to get some. Six weeks I was back to normal. . . . Made a tea out of it."
—Justine James Sr.

Traditional Management and Care

Elders describe the harvest of devil's club, more than that of many plants, as a culturally sensitive practice with special protocols. Efforts are made to not overharvest plants used for medicinal purposes and to help the plant thrive, such as through the strategic pruning of devil's club or neighboring plants. The plant grows very slowly, so cautious harvests are required if it is not to be harmed or overexploited. With so much attention to the medicinal value of devil's club, there is a growing commercial interest in the plant and a risk of overharvest. Some families have their own special patch they return to year after year; any patch showing any signs of past human management should be respected and avoided.

Cautions

Use extreme caution with the spines of devil's club, which are sharp and often hard to remove from skin. Some people have a heightened sensitivity to the spines, and handling them can cause severe swelling. Likewise, use caution with medicines and, as with all such things, consult physicians for guidance if you are unsure. In small doses, devil's club medicines are generally safe, but at higher doses can cause symptoms like accelerated heart rate.

Long clusters of brilliant red berries appear atop densely thorned stalks, encircled by sprawling lobed leaves—each leaf maple-like, but with imposing spiked veins.

Huckleberries and Blueberries

nəḳalčin (evergreen huckleberry)

maniḳınmuut (mountain or black huckleberry)
t̓oˊḳlumnix or c̓eaḳalčin (red huckleberry)
ʔəsuḳalčin (bog blueberry and prob. Cascade bilberry)
sk̓ux̣snil (oval-leaf blueberry)

Vaccinium ovatum (evergreen huckleberry)
V. membranaceum (mountain or black huckleberry)
V. parvifolium (red huckleberry)
V. uliginosum (bog blueberry)
V. deliciosum (Cascade bilberry)
V. ovalifolium (oval-leaf blueberry)

Traditional Uses

Quinault country offers a wealth of huckleberries and wild blueberries, all closely related and delicious. Several species live on traditional Quinault lands and were picked and eaten in huge quantities historically as a delicious staple food—flavorful, nutritious, and culturally important. Though harvests of these berries are not as big as they once were, many families pick and eat the berries every year, as gathering is still an important social and subsistence event.

Each place has its own distinctive huckleberries and wild blueberries. The berries of evergreen or "shot" huckleberry (*Vaccinium ovatum*) are small but tasty, with clusters clinging to the bush later than almost any other berry, well into fall and winter. Reflecting this fact, the Quinault name for this plant, *sK'iuxsnil,* means "winter huckleberry bushes." These berries are available close to the ocean and bays, and often close to settlements, but are uncommon inland. Especially common on the prairies of the Olympic Peninsula are the bog blueberries (*Vaccinium uliginosum*)—small bushes on wet ground, with incredibly sweet and large berries. And, especially where ancestral Quinault managed these patches with fire and other techniques, bog blueberries grow in short, dense patches, and people often have to bend down or even lie on the ground to pick. In higher mountains, small bushes of Cascade bilberry (*Vaccinium deliciosum*) provide berries of comparable size and flavor. On the forest margins and in deeper woods are red huckleberries (*Vaccinium parvifolium*) and oval-leaf blueberries (*Vaccinium ovalifolium*), which are especially eaten fresh, often mixed with other berries to add flavor and texture. Red huckleberries are especially abundant on the partially shaded forest floors of the region, the upright shrubs dotted with pink-to-red berries, rich in vitamin C; gathered in the summer, these berries can also be dried, or mashed into cakes. In recent times, families home-can red huckleberries, make them into jam, or bake them into pies. And, high in the mountains are productive patches of the mountain or "black" huckleberries (*Vaccinium membranaceum*). Intensified through the use of fire and other techniques, patches of these berries were once the focus of major summertime gathering trips to the mountains, and a cornerstone of the traditional Quinault diet and "seasonal round."

"They stayed [at mountain huckleberry camps] for weeks. They made cedar root huckleberry baskets. The berries kept well in damp [storage pits]. They'd be having berries, filling the baskets up before they come down. They had horses up there to carry the baskets down. Potato Hill—they have a campground there. The berries were always free to people that came. They could stay there until they got their berries." —Katherine Barr

All these berries are popular fresh, eaten one by one or by the handful. Evergreen, mountain, and other huckleberries and blueberries are also traditionally dried or smoked, then partly mashed, pressed into cakes, and wrapped in leaves or bark for later use. Another very similar recipe involves mashing the berries, pressing the mixture firmly into baskets, and drying these mixtures near the fire to make loaves of cooked berries. The berry loaves made this way were hugely popular historically. Staying well preserved, they could be consumed throughout the year. Sometimes the berries were also processed without mashing, being smoked by the fire or preserved by submerging berries in containers full of grease. In past times, grease was a popular condiment used on many huckleberry products—added as they were served, if not during their preservation. In recent times, huckleberries and blueberries are canned, mashed and cooked into jams and jellies, or baked into pies. Many people mix their huckleberries and blueberries with other types of berries, especially tart varieties—resulting in rich flavors, complex textures, and the sweetening of the whole mixture.

Huckleberries, red in particular, were harvested with comb-like "berry rakes" that could be run through the branches, plucking off the berries but leaving leaves and stems undamaged. In fact, one Quinault name recorded

Red huckleberries, bright red to pink, can be picked with wooden "berry rakes" that remove berries while leaving delicate leaves unharmed.

Oval-leaf blueberry fruits, looking much like the berries of their domesticated kin, adorn leafy shrubs on the shady forest floor.

for red huckleberry, *to xlumnix,* means "combing off the berries." Baskets of berries gathered in this way often contained many leaves. By rolling the berries down a wet cedar plank, harvesters were able to remove the leaves, which would stick even as the berries rolled into a basket below.

Berry patches are traditionally managed through fire, which kept back the shading forest and released nutrients into the soil. This was done on the prairies of the Olympic Peninsula, as well as in the mountains such as in Enchanted Valley on the upper Quinault River. The burning of huckleberry patches was said to be a spiritual practice, traditionally, with several motives. In part, the practice showed "respects" to the plants and the spiritual forces and beings involved in their productivity. Where these practices were carried out most intensively, one can still find dense remnant patches of huckleberries.

"[People pick blue] huckleberries here, and the shot berries [evergreen huckleberries]. The old people: used to see them out there with their baskets. They'd be climbing up on the old stumps and peeling those shot berries off and boy, everybody'd think, 'Boy, we're going to have shot berry pie, now!' And . . . my dad said that they used to go to the mountains every summer. They'd camp on the upper Quinault River and they'd dry elk meat. And they'd gather—they got them little bushes . . . little bush huckleberries up there. And, boy, those things are good. . . . They're right up next to the ground, they're like a shrub, but there's just acres of them up there."—Francis Rosander

Especially in times before automobiles were common, huckleberry and blueberry harvests were significant social events, with entire families gathering and camping for days while they picked and processed berries in fire-managed patches, in the mountains or along the prairies. Sometimes families from multiple villages or tribes converged on the most productive berry patches to visit and pick. During these events, everyone ate berries and socialized. Women and children were primarily responsible for picking, while men fanned out into the adjacent mountains and valleys to hunt, returning to the berry camps occasionally to dry their meat, visit, and help process berries. At the end of the harvest, these families hauled large quantities of berries home by foot, canoe, or horseback and wagon. They picked and brought home extra to give to the elderly or others who had been unable to participate in the big harvests. By the twentieth century, especially productive pickers sometimes sold their extra huckleberries to tribal members and non-Indians alike.

Huckleberry leaves are also used in making a healthy tea—especially the leaves of red huckleberry, mountain huckleberry, and probably oval-leaf blueberry. A few tribal members also make a tea from the new branches of the evergreen huckleberry, when the leaves are soft and light green in the spring. This huckleberry twig tea is said to cleanse the body, bonding with toxins and unwanted bacteria and carrying these out of the body; regular drinkers of this tea warn that it must be consumed in moderation. There is also some tradition of people smoking evergreen huckleberry—the dried leaves and berries—before tobacco became commonplace.

What They Look Like

Though all huckleberries are related and structurally similar, each huckleberry species has its own distinctive look.

Evergreen huckleberry is a tough upright shrub, often head high, with small leathery leaves—dark, shiny, and slightly serrated. Clusters of small pink bell-like flowers turn to tiny, round shiny black to purple-blue huckleberries—sometimes covered in a whitish waxy coating.

Black or "mountain" huckleberries, gathered summer through early fall in the Olympic Mountains and other high elevation plots, are a traditional staple of Quinault and many other Northwest tribes.

Mountain huckleberry (black huckleberry) is a small shrub—often knee- to waist-high—with thin, oval leaves with pointed tips. Pinkish urn-shaped flowers turn to medium-sized shiny purple-black berries.

Red huckleberry is a large, loosely spreading shrub with small rounded light-green leaves. Greenish urn-shaped flowers turn to pink to deep-red berries hanging closely to the undersides of branches.

Bog blueberry is a low, often creeping bush with twiggy stems and smooth, oval leaves with visible veins. Pink urn-shaped flowers turn to sweet blue berries, often oblong, with whitish waxy coating.

Oval-leaf blueberry is a lanky shrub with reddish-brown stems and relatively large light-green oval leaves. White to pink urn-shaped flowers hang closely from stems below the leaves. Individual berries look much like commercial blueberries, but are borne singly on branches.

"My wife every year makes a couple of wild blackberry pies with the huckleberries mixed 50-50. The huckleberries are so sweet and the tartness of that wild blackberry that's just awesome. You don't get a better pie. Makes a couple of them every year. You just can't beat it.... It's awesome. You know how sweet shot berry is, and then the tartness of that wild blackberry: [it's] the combination of the two." —Gerald Ellis

Where to Find Them

Evergreen huckleberry is found in low-elevation coniferous forests, from coastal beaches to the foot of the

Olympics—especially in well-lit forest clearings and edges. Mountain huckleberry is found in well-drained mountain forest clearings and some open slopes at middle to high elevations—especially where ancestral Quinault maintained patches through the use of fire and other techniques. Bog blueberry is especially found in the prairies of the Olympic Peninsula, and in other damp places with acidic soil; this berry can also be found in moist subalpine meadows in mountains and hills now on US Forest Service and National Park Service lands. Red huckleberry is found in open forest and forest clearings from low to middle elevations, often growing on stumps or downed wood. Oval-leaf blueberry grows in similar conditions, but often in denser, shadier coastal forests.

"The prairies [had] a lot of blueberries. A lot of winter huckles . . . nice big berries. . . . We used to pick them up in Moses Prairie. Oh, they're sweet, but they're little! Just prairie blueberries. Hell, we used to take a blanket, and lay down and eat the things. A lot of times, you're waiting for an elk to move around, so you just lay down [and eat] . . . on a small bush like that—it wasn't fifteen, eighteen inches high. Some of the best berries you could ever find. . . . There were a lot of them . . . all throughout the prairie. We never got a hundred yards off the road. Go way down in a patch and eat berries."—Justine James Sr.

When to Gather

Prime huckleberry picking time is in the summer, with some species and patches extending into the early fall. Evergreen huckleberries stay on the bush later than most other species and are sometimes available into fall or even early winter. When making medicine, such as with leaves or evergreen huckleberry stems, spring is the usual gathering time.

Traditional Management and Care

Burning was essential to traditional Quinault maintenance of many berry patches, especially those in the prairies and in the mountains. Fire kept away trees that might shade and crowd out berry patches and released nutrients to the soil, allowing huckleberries to expand and become the dominant plant in some of the more popular picking areas. Other methods, including the scattering of berries at traditional harvest sites, may also have been used. If one encounters large patches of mountain or bog huckleberry in Quinault country today, there is a good chance the patches were maintained by the ancestors. In recent times, a few tribal members have successfully transplanted berry bushes to their yard; success is greater if the bush can be planted in conditions (sunlight, moisture, soil) similar to the plant's original home.

Dusty blue bog blueberries, thriving in damp places, are a small but delicious part of the traditional Quinault diet.

Evergreen huckleberry grows on robust shrubs, with clusters of very dark purple berries that remain on the bush, improving in flavor well into the late autumn.

Salmonberry

k̓ʷəla or smətčə | *Rubus spectabilis*

Traditional Uses

Salmonberry is among the most important food sources in the traditional Quinault diet. This plant's fresh shoots are the favorite spring sprout of many Quinault families, who peel the prickly outer skin and eat the soft, sweet inner stem. These sprouts are widely appreciated as a healthy, nutrient-rich plant food that helps people in their transition from winter to the active summer season. They are also said to provide a healthy balance with the traditional diet of the Quinault, so rich in fish, shellfish, and meat. People often pick and eat large quantities of these sprouts as they travel, or bring them home for a meal the day they are picked. The sprouts are described as a popular snack for children playing outdoors or people out hunting and fishing in the springtime.

Beyond eating the sprouts fresh, Quinault families have various recipes and ways of processing and preserving them for later use. A few Quinault people gather and refrigerate the sprouts, though they last only a week or two refrigerated. Elders also report steaming peeled sprouts or even simmering whole spouts, leaves and all, then peeling the stalks when the sprouts cool. Traditionally, these sprouts are dipped in oil and eaten fresh or steamed in pits—with oil of seal or blueback salmon (sockeye, *Oncorhynchus nerka*) especially popular. In Quinault cuisine, steamed salmonberry sprouts were often eaten along with dried smoked salmon. The sprouts were also dipped in "stink eggs," a mixture of salmon or steelhead eggs that were buried and allowed to ferment underground. As Gerald Ellis recalls, "When the salmonberry sprouts get ripe, you dug them up and dipped your sprouts in there and ate it. By right, they were called stink eggs!" The light flavor of the salmonberry sprouts was said to offset the heaviness of the fermented eggs, and the oily eggs helped relieve constipation, which was experienced by those who ate a lot of sprouts. Together the whole egg and sprout mixture was said to "clean them out" after a long winter spent immobile and indoors. The very young leaves of salmonberry were also sometimes eaten, often cooked, and dipped into oil or stink eggs—like salad or steamed greens.

"Oh, salmonberry sprouts.... The Indian word for sprouts is ja'ns. *... They're really good. I fished up the river there for years—when they were ready to eat, I always bring a bunch home, put them in the refrigerator, they last for a couple of weeks."*
—Clifford "Soup" Corwin

Salmonberries—with colors ranging from yellow-orange to deep reddish-purple—are themselves a keystone of the traditional Quinault diet. These berries are gathered in large quantities and usually eaten fresh. The berries are also made into jellies or jams, pies, and other foods. Some families also have canned salmonberries in jars—often mixed with other berries, like red elderberry, that add to salmonberry's subtle colors, flavors, and textures. The berries are so moist that they were not easily dried into traditional cakes or fruit leathers, as was done with other berries. Instead, salmonberries were placed in bentwood boxes or other containers, and oil was poured over the berries to seal and preserve them for later use. When the berries were pulled from the container, some of the oil was squeezed out and some remained as a condiment.

Even the woody older sticks and stems of salmonberry are valuable. These were scraped of thorns and used as roasting skewers or to suspend drying clams for air-drying or smoking. For hunting small game such as rabbits or birds, Quinault people sometimes constructed salmonberry-shaft arrows, or double-pointed arrows, their points made of woody salmonberry stalks, tips hardened by heating over a fire. Especially thick and straight salmonberry stalks, similarly processed, sometimes served as light-duty salmon spears. Thinner salmonberry sticks, pushed through the round upper roots of skunk cabbage, functioned as the center pin of spinning tops used in traditional games (see skunk cabbage). Another game involved fashioning small salmonberry-cane spears, with competitors throwing them at targets made of fungi—probably bracket fungus—to see who had the best aim. The bark of these woody canes also has medicinal value. The bark of the mature stalks, when peeled, was simmered to produce a decoction used to remedy infected cuts and burns. Furthermore, women traditionally consumed a hot

Baskets of freshly picked salmonberries show the full spectrum of berry colors found in a single thicket, from yellowish-orange to deep purplish-red.

infusion of salmonberry bark as a tonic before and after menstruation; and, boiled in seawater, the decoction has been ingested to lessen labor pains.

Reflecting the importance of this plant, many Quinault story cycles mention salmonberry, with pivotal events often centering on the experiences of women picking sprouts or berries. There are stories of springtime guests being frustrated by how their host, Beaver—so fond of woody food—tries to serve them old, hard salmonberry stalks, passing them off as tender berry sprouts. Accounts from ancient times mention the "salmonberry bird," Swainson's thrush, singing its song to ripen the berries, helping to feed all living beings. That song, heard in the woods even now, still heralds the arrival of salmonberries in Quinault country. Salmonberry blossoms are known to appear at the same time that the blueback salmon are returning to the Quinault. As the berries ripened, people historically returned to fish in the Taholah area from social, ceremonial, and resource-gathering trips up and down the Washington coast. Each year, many returned from villages as far away as the lower Columbia River, where the Chinook also marked the correspondence of the spring salmon return and that of the berries—the berries playing an important role in Chinookan first salmon

"Arrows, actually, some of the best arrows that we made were from the salmonberry bush, once they, you know, age and become brown. Pick them. Seem like they break really easy, really weak, but they're a perfect weight and density for swiftly flying through the air. You got to soak them to straighten them out. You have to—what's it called—temper it, put it by the fire and dry it out, to make sure you keep it straight."
—Micah Masten

ceremonies. This spiritually potent natural choreography impressed early non-Indian explorers on the Columbia tidewater. Through their written accounts of this culturally significant place and practice, the plant's name, "salmonberry," came into global usage.

What It Looks Like

Salmonberry is a rangy, tall, prickly shrub with sharply toothed, usually three-part leaves, like raspberry leaves. Five-petaled simple pink to magenta flowers turn to complex raspberry-shaped berries with colors ranging from yellow-orange to deep reddish-purple; berry colors are the same on any one bush.

Where to Find It

Salmonberry is found in moist areas such as on the banks of streams, wetlands, and in recently disturbed damp forest from low to subalpine elevations, the best patches often forming dense thickets.

When to Gather

Berries arrive in late spring and early summer, ripening earlier in warmer places and at lower elevations—their arrival loosely correlates with the return of the blueback salmon. The edible berry shoots emerge in the spring, but late shoots can still be gathered into the summer in places where sprouting is delayed, such as in the mountains. Edible shoots are soft, bendable, and easy to break off, but become firm and woody and inedible into the summer. Leaves are available in the spring and summer months.

Traditional Management and Care

In some settings, such as on prairie and meadow margins, fire appears to have allowed salmonberry patches to thrive and expand. In some places, people intentionally scattered the seeds from processed salmonberries, and even placed fish waste or other materials on these patches, which enhanced their growth. As Chris Morganroth recalls, "Wherever they were strewn, they just kind of took over. They grow fast and if they take care of them, like putting in a fertilizer, salmonberries can grow [well]. Thousands of more berries if they were fertilized."

Cautions

It has been written that people who ate very large quantities of salmonberry sprouts have sometimes been reported to have skin outbreaks, or swollen eyes—to the point that some peoples' eyes swell nearly shut. The effect was said to be temporary. The reliability of these claims is unclear, as this effect is seldom seen today.

Maggie James Kelly, a Quinault woman, picking salmonberries in traditional cedar bark clothing, circa 1912 (upper left). An edible salmonberry shoot, peeled (upper right). Magenta salmonberry blossoms (lower left). A very ripe orange salmonberry fruit (lower right).

1368.
1.
J. L. delt.
Pub by J. Ridgway 169 Piccadilly Nov. 1. 1830.
J. Watts sc.

Thimbleberry

hiʔinis | *Rubus parviflorus*

Traditional Uses

The brilliant red berries of thimbleberry are popular wherever they are found. At certain times, in certain patches, the ripe berries are sweet and slightly crunchy with many small seeds. In some places they are slightly bitter, but as Phil Martin Sr. said of thimbleberries, "Oh, they're bitter, but our people loved bitter! Old timers seemed to love bitter." Sweeter berries are often eaten fresh, while bitter ones are mixed into jams, jellies, pies, and other foods. These berries are patchy but abundant on parts of the Quinault coast, and are also found in the mountains and gathered as people traveled historically. If patches of unripe berries were encountered, people picked these in large quantities and stored them in baskets, often lined with thimbleberry leaves. In time, the stored berries ripen off the plants in containers and can be consumed when ready. Even relatively ripe berries are sometimes said to improve and sweeten when they sit in such baskets for a time.

The fresh, slightly bendy shoots that grow in the spring are also popular as food. Many of these juicy shoots are sweet and fruity, with Quinault people peeling the outside and consuming the soft inner stem. These can be eaten fresh or steamed like asparagus, and are traditionally dipped in oil. They do not store well, so are typically a springtime food only—a healthy, nutrient-rich food that helps people transition from winter to the active summer months.

The large, soft leaves of thimbleberry also have important uses. Baskets of elderberries, thimbleberries, and perhaps other species are traditionally lined with thimbleberry leaves. This protected the baskets from staining or sticking to the berries, but also helped "cure" the berries. Thimbleberry leaves are also the traditional wrap for certain foods, such as clover roots (from the springbank clover, *Trifolium wormskioldii*) or, less commonly, white-fleshed fish when they are steamed or roasted in pits. This gives bland foods a very subtle fruity flavor. The fresh leaves are occasionally used in traditional teas and tonics. Crushed leaves are sometimes used in poultices and can help reduce the sting, for example, of stinging nettle.

Bright white thimbleberry blossoms with five petals each, perched atop soft green stalks, resemble those of their distant wild rose kin.

What It Looks Like

The large, fuzzy, long-stalked light-green maple-like leaves of thimbleberry extend from woody stems, growing up to eight feet tall. Large attractive white flowers —five-petaled and wild-rose-like—turn to red domed berries somewhat resembling flat raspberries.

Where to Find It

Thimbleberry grows in clearings and open woods at low elevations, often in large patches, being especially productive in moist soils. It can occasionally be found at medium elevations along stream corridors.

When to Gather

Berries appear in late spring and early summer, ripening earlier in warmer places. Berry shoots are gathered in the spring, but late-sprouting shoots can still sometimes be gathered into the summer, in some places such as in the mountains. Edible shoots are those that are bendable and soft; they grow firm and woody through the summer and are no longer edible. Leaves are gathered in the spring and summer months.

"Whenever my mother went out, I was always carrying buckets and jars and things to carry berries in. We'd can them and freeze them and use them to make pies and freeze the pies. We did a lot of that. Especially salmonberries, thimbleberries."
—Marlene Hanson

Traditional Management and Care

In some settings, such as on prairie and meadow margins, fire appears to have produced the clearings that allowed thimbleberry patches to thrive and expand.

Cautions

Thimbleberries are one of the berries traditionally avoided by pregnant women, as it was said that they would cause infants to develop birthmarks. This may relate to the fact that the plant's leaves, like raspberry leaves, are reported to have some effect on women's reproductive cycles and may cause uterine contractions. Pregnant women may wish to consult with a knowledgeable physician before consuming thimbleberry products in large quantities.

Relatively flat, deep red berries, sweet, with crunchy seeds, emerge to replace white thimbleberry flowers in spring and summer.

Light green and delicate, an edible young thimbleberry shoot emerges from within a patch of salal.

Blackcaps, or “Black Raspberry”

t̓s′lekoma | *Rubus leucodermis*

Traditional Uses

The very popular but hard-to-find berries of blackcaps are like a dark raspberry in flavor and appearance. They are a favorite eaten fresh off the vine—so flavorful and uncommon that people eat them right away, sharing them with family and, traditionally, mixing the berries with oil. A lot of impromptu berry picking and eating seems to have occurred as people traveled inland to fish or to hunt and gather berries in the Olympic Mountain foothills. Still, blackcaps are occasionally preserved, mixed with other berries, and cooked. Berry “cakes” sometimes included blackcaps, pulverized, dried, and pressed into rounds for later use. Blackcap fruit leather can also be made by spreading pulverized berries on skunk cabbage or other leaves for drying; and, in recent times, the berries have been used in jams, jellies, pies, and baked goods. As with other berries of its genus (such as thimbleberry and salmonberry), blackcap shoots in the spring are sweet, fruity, and traditionally consumed with oil. The leaves may have been used in medicines, such as in teas for digestive complaints.

What It Looks Like

Blackcaps are a sprawling to upright vine with bending stems; the numerous thorns and sharp-toothed leaves, usually in sets of three, are white on the undersides. White flowers like tiny roses turn to deep purple-black raspberries.

Where to Find It

Blackcaps are found in small patches in forested areas with dappled sunlight on the forest floor, especially on low hills and lower slopes of coastal mountain ranges. Quinault ancestors historically burned over areas to

"We have a lot of black-caps. They're really good. The story is, there's this [hill]. . . . Used to be a real good fishing set there on the river. And there was kind of a mountain up behind and we used to say, 'Yeah, my grandfather told me that was a little hill when he was a boy.' Because, you know, it was pretty good sized. On the back side of it was just, the whole thing was just blackcaps. People used to come up there and pick those."
—Francis Rosander

enhance native vegetation, producing highly productive blackcap patches. Many people have their favorite family patches that they return to year after year.

When to Gather

The edible new shoots of blackcaps appear in the spring, while the berries ripen in early- to midsummer, depending on elevation.

Traditional Management and Care

Blackcaps are clearly concentrated in forested areas that were traditionally managed through the use of fire, such as Burnt Hill, and appear to be among the berry species Quinault people sought to enhance at those places. Elders report that some have tried to transplant blackcaps, with varying results.

Cautions

Traditionally, blackcap was one of the berries avoided by women during pregnancy; the consumption of this berry was believed to increase the likelihood of birthmarks on the baby. As with thimbleberry, consumption of this plant has the potential to affect pregnant women and should be done with caution and the input of medical professionals.

Blackcap berries are red to reddish-purple when they first appear, but darken to shades of deep purple and purple-black as they turn ripe and sweet.

Serrated bright green leaflets, deeply veined and growing in groups of five, surround blossoms of five delicate petals, their colors ranging from white to pink.

Blackberries

waʔs (native trailing blackberry)	*Rubus ursinus* (native trailing blackberry)
šwaha (other blackberries)	*R. armeniacus* (Himalayan blackberry)
	R. laciniatus (evergreen blackberry)

Traditional Uses

Among the berries gathered traditionally, blackberry is among the tastiest and best known. Historically, Quinault families have gathered the native or trailing blackberry (*Rubus ursinus*), with its sweet, small berries growing near the ground on long, thin, vine-like stems. These berries are still gathered in many parts of Quinault country, along roads, on the edges of settlements, with their lanky stems winding through the branches of bushes and low trees. These berries can be eaten one by one straight off the vine. Traditionally they were also popular as the main ingredient of berry "cakes" made from mashed blackberries, pressed into baskets and dried near the fire. The preserved cakes were hugely popular and were consumed throughout the year.

The berry of the wild blackberry is smaller and more oblong than its domesticated and naturalized kin, with a distinctively sweet fruity flavor preferred by many elders.

As non-Indians moved into the area in the nineteenth century, the available kinds of blackberries changed. Settlers introduced Himalayan and evergreen blackberries, two species previously unknown in Quinault country—a few planted intentionally, the rest propagated as birds spread seeds widely to meadows, farmed fields, roadsides, and clear-cuts throughout the Pacific Northwest. These berries were new, but looked and tasted familiar, and Quinault families happily began picking them alongside the native trailing blackberry harvest. In spite of the similar look and taste of these introduced berries, they have a noticeably different flavor. Even today, Quinault elders express a strong preference for the native trailing blackberries

Families report going to prime picking areas and gathering huge amounts of blackberries. Some have regularly gathered whole washtubs of berries in a single picking.

Many of these berries are eaten fresh or promptly made into jam, jelly, or pies. Yet, with such large quantities, families are able to freeze, can, or otherwise preserve blackberries, so they can make pies and other blackberry-filled treats throughout the year.

Blackberries are popular with nearly everyone. Accordingly, some Quinault families have been able to pick and sell blackberries for extra income. In many off-reservation places, development has overtaken and eliminated some families' favorite berry patches. In some cases, Quinault families have been able to arrange for continued access to these private lands, sometimes bartering a few berries or other plant foods for continued access.

"The little ones, the little wild ones, they're the ones that are really good. . . . Them and Himalayas are the same, only the Himalayas are a bigger berry. The evergreens is a tighter berry. The little ones are called wild blackberries. They grow on vines on the ground. I used to go up to the mountains and pick them. In them logged off areas. I'd can twelve to fourteen cases of them a year." —Justine James Sr.

While the berries of blackberry plants are the main attraction, other parts of the plant may be used as well. A few families eat blackberry shoots. While the thin berry shoots of the native trailing blackberry are infrequently used, families living close to Himalayan blackberry patches have sometimes picked, peeled, and eaten shoots from those plants like they would salmonberry sprouts. Blackberries also have medicinal uses. A decoction made from the leaves or roots is sometimes used as a diarrhea cure. The berries are also said to help with related stomach complaints.

What They Look Like

Native trailing blackberry has long, creeping vines of a blue-green to gray color, lined with small thorns. The leaves are usually three-parted, the leaflets pointed, toothed, and deeply veined. White flowers, usually with five petals, turn to firm, elongated purple-black berries.

"The blackberry root [for] diarrhea cure . . . you just boil it up and drink it like tea. Of course you have to clean all the dirt off it!" —Lucille Quilt

In contrast, *Himalayan blackberry* has relatively tall, arching reddish-brown to green canes, rounder leaves often with five leaflets, and larger flowers and fruits. *Evergreen blackberry* is also larger, each leaf with five leaflets—the most noticeable difference being that evergreen blackberry leaves are relatively complex, with leaflets sharply toothed with deeply cut lobes, to the point of giving them a "lacy" appearance. The petals of the evergreen blackberry flower are lobed, also giving the blossoms a lacy appearance.

Where to Find Them

Trailing blackberry grows in open woods and in clearings such as clear-cuts, along streams and roadsides. The introduced Himalayan and evergreen blackberries are especially abundant in disturbed sites, such as along roadsides and on the edges of settlements and on vacant lots in urbanizing portions of the Olympic Peninsula.

When to Gather

Berries appear through the summer, especially in mid- to late summer—the final berries of the season often begin to rot soon after the first fall rains. Edible shoots are available in spring. Roots for medicinal purposes can be gathered as needed but are most commonly gathered in the early spring or fall.

Traditional Management and Care

Blackberry is among the plants that rebound well at the edge of burned clearings, and Quinault ancestors once enhanced growth through the use of fire. At the same time, fire has been used to keep nonnative Himalayan and evergreen blackberries from taking over meadows of the smaller, less aggressive native trailing blackberry. The privatization of land and development in places like Aberdeen and Ocean Shores has reduced or eliminated access to many of the larger blackberry patches.

Growing in long, creeping brambles close to the ground, wild blackberry has white flowers of five or more petals, each petal long and delicately thin.

Elderberries

k̓əluʔm (red elderberry)
k̓weʔlap (blue elderberry)

Sambucus racemosa (red elderberry)
S. caerulea (blue elderberry)

Traditional Uses

Throughout the summer months, tiny clusters of elderberries, red or dusty blue depending on the species, are seen on tall bushes throughout Quinault country. In various ways, these berries have played an important role in the traditional ethnobotanical practices of Quinault people. Elderberries receive mention in ancient oral traditions that describe religious specialists helping to ensure good elderberry harvests and beings who refuse to eat elderberries sometimes going hungry.

Available in large quantities in the summer, with red ripening before blue, the berries have been gathered and stored by Quinault people for use in the winter months—blue in particular. The berries were traditionally steamed in pits dug into the sand and lined with rocks. In this process, a fire is set in the pit, and once the fire burns down, alder bark is placed inside the pit and covered with skunk cabbage leaves. Next, the elderberries are wrapped in skunk cabbage or other leaves and placed in the hot pit. The pit is then doused with water to produce steam, buried, and the berries are allowed to heat for many hours. When the pit is re-excavated, Quinaults traditionally crush the berries before placing the mixture in boxes of hemlock bark or other materials, covering the top of the baskets in leaves of vine maple or other species and submerging whole baskets of cooked elderberries in excavated cold-water trenches for winter use.

Raw red elderberries are widely reported to be poisonous, largely due to toxic compounds in the seeds; the green parts of both red and blue elderberry plants also contain toxic compounds, whether raw or cooked, and consuming them in any form is not recommended. Many Quinault elders, however, eat the raw berries enthusiastically, saying one builds tolerance and a taste for the berries over time. Safe processing for food use requires special knowledge and care. Cooking dissipates toxic compounds in the berries, and those who gather the berries today commonly remove the seeds. Red elderberries are commonly mixed as a secondary ingredient in mixtures of other berries, such as salmonberry, to add color, texture, and flavor. Families that can berries such as salmonberry in jars often add red elderberries for this purpose.

"We have the red elderberry out here. My grandmother gathered those up like crazy. She would can them or preserve them somehow. People would say, "They're poisonous." I'd say, "Naw, they're not poisonous. They're 'slightly toxinous!'" Once you get used to eating them, then they don't affect you. You can eat a little bit of it and later on eat a whole bunch and you're pretty much used to whatever it is that they do to your body."
—Chris Morganroth

While uncommon on the outer coast, blue elderberries are sweeter, more popular, and eaten wherever they are found. Some people eat the berries raw in the forest, while others gather them and bring them home for later use. Elders like Katherine Barr recall her family's tradition of adding a little sugar to blue elderberries and simmering and eating the berries like they did applesauce. Tribal members report happy experimentation with blue elderberry for many uses: pies, jam and jellies, baked goods, even wine.

Elderberries have played a role in Quinault medicine as well. The bark has often been used to fight infections in cuts and other wounds. People scrape the elderberry bark from thicker stalks, simmer the bark, and soak a cloth in the decoction; this cloth is then placed on wounds to reduce or prevent infection. Elders note that wounded animals rub against elderberry bushes for the same effect. A decoction of elderberry bark or root was sometimes placed on new mothers' breasts to bring in their milk. Elderberry infusions are also taken internally, though conventional sources suggest that this brings risks of toxicity. Drinking a mild decoction from the bark or roots has been said to help with diarrhea, sore throats and colds, and was sometimes used as a tonic for women during the later stages of pregnancy. Making use of its toxic properties, a strong decoction of the bark or roots was used as an emetic, causing vomiting for the purpose of cleansing, to expel poison, or to help "reset" a chronically upset digestive tract. A few elders mention chewing a few fresh new sprouts from the elderberry bushes as a hunger suppressant when hiking or canoeing in the spring.

Elderberry stalks, being largely hollow, with interior pith that is easily removed, served many purposes. With pith removed from the stalk and one end plugged, the stalk could be fashioned into an elk whistle. With one end plugged, the stalks also served as containers for small objects such as beads; or as straws, snorkels, or "pea shooters"; and for many other purposes. The elderberry makes a good, long-lasting dye for certain materials—red or purple from the berries, green from the foliage, or black from the charred roots.

What They Look Like

Red elderberry is a large, rangy shrub that reaches 20 feet, with arching branches. Compound leaves have 5 to 9 lance-shaped leaflets with jagged edges. The small creamy-white flowers grow in pyramidal clusters, ripening to dense bunches of tiny bright-red berries.

Blue elderberry is a denser and sometimes taller bush, up to 30 feet high, also with compound leaves and leaflets that are pointed and somewhat curled compared with red elderberry. The berries, blue, and covered with a powdery coating when ripe, are borne in dense flattened or rounded clusters.

Where to Find Them

Red elderberry is shade tolerant but thrives in full sun on moist soil near streams and swampy areas from sea level to middle elevations; it can be found widely on the Olympic coast, on the margins of older forest, prairies, and moderately lit clearings. Blue elderberry is more common inland, from low to middle elevation, in moist clearings.

When to Gather

Elderberries are gathered in the summer—July being the peak month for red elderberry gathering traditionally, while blue elderberries appear only in late summer in the Olympic Peninsula. Stalks for making traditional crafts can be gathered any time, while bark for medicinal purposes is most commonly harvested in the spring. Roots are especially potent when harvested in the fall through winter.

Drooping clusters of tasty, powdery-blue berries, ready to pick from a summertime blue elderberry bush.

Traditional Management and Care

Traditionally, fire probably expanded the range of elderberry on the margins of prairies and forest clearings. Good blue elderberry patches are relatively few, many being close to old Quinault settlements; they may have been enhanced by various means, including selective berry harvests. Plants that are used medicinally, for elk whistles, for dye, and other important purposes, are often shown proper respects, including efforts to not kill the entire plant.

Cautions

Elderberry leaves, bark, roots, and seeds contain cyanogenic glycosides—cyanide-producing compounds that can be very toxic. For this reason, eating too many elderberries raw can cause nausea and vomiting. People have different levels of sensitivity; cooking and other processing can reduce the toxicity of the berries somewhat. Still, some sources insist that consumption could involve risks to sensitive individuals; therefore, any internal use of red elderberry or of any elderberry bark or leaves, should be undertaken only with the guidance of a medical professional.

2845.
1.
2.

Salal

k̓waʔsoičin | *Gaultheria shallon*

Traditional Uses

Salal is both widespread and widely used in Quinault country. First introduced to the non-Native world by Chinookan traders at the Columbia River's mouth, both the common name "salal" and the scientific species name shallon come from the Chinook word for the plant. Some Quinault, including many of Chinook ancestry, call these berries "bear berries" today.

The berries of salal are very popular—eaten raw, but also canned or cooked. Traditionally, Quinault people mash the berries, dry them, and press them into "cakes" or spread them on cedar planks or skunk cabbage leaves to make fruit leather. Traditional recipes varied, but berries were often heated or smoked on a mat of fern fronds placed over a cedar lattice a few feet above a fire. Once partially cooked, the berries were mashed, molded into "cakes" and dried again until ready for storage. To eat, a portion was sliced from the berry cake or leather, then soaked in water or dipped in whale or seal oil. Some accounts say the Quinault submerged watertight baskets of such pulverized salal berry foods in refrigerating ditches filled with cold spring water, preserving them for later use. Meanwhile, salal leaves were commonly used to line cooking pits or spread out and used as a surface for drying foods. Salal branches covered in leaves are often the preferred material for lining pits for cooking camas, with the leaves imparting a more pleasant flavor to the edible bulbs than many other leafy species.

"They pick a lot of them—even can them. They're good, canned [but] they can be right off the vine—when they're fully ripe, they're good!"
—Katherine Barr

Today, tribal members have many new recipes, making delicious jams, jellies, and syrups from salal berries. A few tribal members make pies, or still mix salal with other berries to make traditional fruit leathers or berry cakes. Some happily eat the slightly tough, "hairy" outer skin of the berry, while others say they peel off that layer or pulverize the berries, such as in a blender, to even out the texture before use in recipes.

*"Salal berries (*skwasa utc n*) were prepared for winter use by smoke-drying. Cedar was split into thin, narrow strips and woven into an open checker weave, and this frame placed on an ordinary drying rack about four feet above the fire. The frame was now covered with fern leaves and berries spread on top. Three or four baskets of berries were placed on one such frame. The drying process was kept up for three days. Then the berries were mashed, molded into round cakes six inches in diameter and these again dried. If the berries were wanted for late winter use, an alder dish large enough to hold five cakes was made and the half-dried berries were placed in this and mashed with a stone pestle. The fruit was now dried still more and stored away. When wanted for use, slices were cut from the cake, mixed with water and eaten with oil."*
—Ronald Olson, 1936

Beyond its food value, salal has many other uses. Salal appears to have been used as a dye for certain plants used in basketry, such as dune grass (*Leymus mollis*). Elders also mention the berries are popular with wildlife, including bears and birds. Meanwhile, salal leaves have many traditional medicinal and ceremonial uses, and also, often mixed with tobacco, were formerly smoked.

Today, the leaves of salal are still widely used in medicine. Some use the leaves as a laxative. As Chris Morganroth recalls, "They would chew the younger leaves that have almost reached its full stage of growth. The leaves would be used just by chewing it, then they would swallow the spit and spit out the leaf. That would serve as a laxative." Others report chewing salal leaves for colic, heartburn, gas, and similar complaints. Boiled lightly, these leaves produce a stomach-settling tea, sometimes used as a preventive against sickness. Boiled until the water goes dark, the leaves are reported to produce a diarrhea tonic. Mixed with alcohol, a dark, slow-cooked salal leaf liquor can become a potent purgative that cleanses the body with sweat and even vomiting. Salal roots and bark are sometimes used in these preparations as well. Salal leaves are widely reported to be used as a powerful antiseptic for cuts and skin infections. Chewed or ground and placed on wounds, some Quinault report rapid healing and reduced scarring from this application. A few tribal members still keep a few salal leaves in their home for these purposes.

Finally, the angular branches and shiny leaves of salal are very popular with the floral industry, and greenery is shipped to cities in the Northwest, elsewhere in North America, and often Europe. Sensitive salal harvests have sometimes been permitted by the Quinault Indian Nation, and tribal members occasionally have made extra income participating in such harvests. Illegal poaching of salal greenery on the reservation and beyond is a growing problem.

What It Looks Like

Salal is an erect evergreen shrub with leathery leaves and hairy stems and branches. Urn-shaped pink-to-white flowers borne along reddish-brown stems turn to dark-blue berries, each with a deep X-shaped indentation on its bottom end.

Where to Find It

Salal forms locally dense thickets along roadsides, in brushy areas on hillslopes and close to the ocean, and on prairie margins. The plant thrives widely, though, tolerating both sun and shade as well as dry to wet soils in coastal coniferous forests, on rocky bluffs, and along boggy areas from low to middle elevations.

When to Gather

Salal leaves for medicinal use are often gathered in summer, when they are still slightly pliable. Berries are especially abundant in mid- to late summer, when they can be picked individually or in long clusters attached to the stem.

Deep purple-blue berries protrude from robust evergreen salal bushes on bright magenta stems.

Traditional Management and Care

Elders report that salal grows especially well in places that have been managed by burning, such as on the margins of prairies. Burning removed old, dead woody growth and allowed for abundant new leaves and berries. Some tribal members still use related pruning techniques to promote the luxuriant growth of plants gathered for personal use or as commercial greens. Those who gather salal leaves or branches for medicinal or other cultural purposes sometimes report showing respects to the plant upon harvesting, or taking steps to ensure the well-being of plants used in this way.

Wild Currants and Gooseberries

łeʔimk̓s (gooseberry—coastal black and others)
p̓ookwa (trailing black currant, stink currant)

Ribes spp.

Traditional Uses

Many species of wild currants and gooseberries exist in Western Washington, all of them related, though the Quinault and other Washington Pacific coast tribes did not eat the berries as often as tribes in other parts of the West. In fact, writers have sometimes claimed that Quinaults never touched these berries. This, however, is not true. Some elders recall their families gathering these varieties, including stink currant (*Ribes bracteosum*), and eating the berries fresh. When ripe, and slightly soft to the touch, the berries are often sweet and can be eaten alone or mixed with other types of berries. A few have preserved these types of berries for later use, though the small scale of the harvest makes this practice uncommon. When it was done, preservation was accomplished by drying the berries, often combining them with elderberries or other kinds of berries and pressing the mixture into cakes. Such preserved berries were formerly kept in underground pits, or in trenches filled by cold spring water. More recently, families have made jams or pies from wild currants or gooseberries, as well as domesticated gooseberries planted in their yards during the twentieth century.

Especially when crushed, the leaves of currant and gooseberry species have strong and distinctive smells—from pleasant and spicy to acrid and nasty. Because their smell can mask other odors, Quinault people have used certain types of currant leaves as air fresheners. In cases where extra attention is needed, such as after a person has died in a house, branches from currant bushes are traditionally spread around the house to help cleanse the space.

"There used to be a gooseberry bush down on the river when I was a little girl. I'd go check that bush every day when they were growing. Every single day. They were red. Then they'd get darker if nobody ate them . . . oh, they were good. I'd go check that thing every day, and start eating on it. Every day. I never took any home—I'd just stand there by the bush and eat it. Go back the next day and have some more!"
—Alicia Figg

What They Look Like

Currants and gooseberries are low-growing woody deciduous shrubs with lobed, maple- to mitt-like leaves. Fuchsia-like flowers, whitish to purple, hang in clusters below the branches. On gooseberries, these flowers turn to clusters of translucent green (or, on coastal black gooseberry and sticky gooseberry, deep purple) berries with visible veins. On red-flowering currant, magenta flower-clusters turn to clusters of powder-blue berries. On stink currant, elongated clusters of flowers turn to powdery blue-gray berries. Gooseberry stems are spiny or prickly, whereas currants are unarmed.

"When my grandmother introduced me to them it was hard for me to stay away from them. I really liked them."
—Chris Morganroth

Where to Find Them

Gooseberries and currants usually grow in moist ground, especially along rivers, streams, and wetland margins, often some distance inland from the sea. Quinault harvesters also report traditions of picking gooseberries and wild currants along the coast as far south as the Columbia River, harvesting concurrent with fishing trips to those areas.

When to Gather

Currants and gooseberries are usually harvested in the summer. Greenery is available much of the year, except winter.

Traditional Management and Care

Currants and gooseberries are sometimes sparse and often grow slowly, so the harvest of leaves and branches is done sparingly. Plants harvested for cleansing are shown proper respects. In some settings, it appears that Quinault people may have enhanced currant and gooseberry production through burning, or even through intentional seeding or transplanting.

The intensely colorful blossoms of red-flowering currant (*R. sanguineum*) are among the most brilliant to be found on the springtime coast, even as its bland berries are mediocre fare.

The black, slightly "hairy" berries of coastal black gooseberry (*R. divaricatum*) emerge from small, fuchsia-like flowers and are gathered where available in the dense coastal forests of Quinault country.

1425.

Oregon Grape

t̓səut́x̣isən

Mahonia aquifolium (tall Oregon grape)
M. nervosa (dwarf or dull-leaf Oregon grape)

Traditional Uses

Common in the forests and mountains of Quinault country, Oregon grape—both the tall Oregon grape and the dwarf or dull-leaf Oregon grape—are showy plants of much value. Most importantly, they are a popular source of dye, which is extracted from the plant's roots. The dye is not just yellow—it is, as Harvest Moon exclaims, "bright, bright yellow!" As such, Oregon grape is the main source of yellow dye in the traditional Quinault tool kit, used for basket materials, regalia, and other purposes. The roots are usually crushed and boiled, and materials such as basketry grasses are soaked in the bright yellow decoction. Paints made with Oregon grape infusions are also reported, mixing deep yellow extract with fats and natural fixatives like salmon eggs.

In addition, Oregon grape is widely regarded as a medicinal plant. A bitter decoction from the boiled roots and rhizomes has been a traditional Quinault medicine for coughs, stomach complaints, and other purposes. Some elders report the use of Oregon grape extract as a "blood purifier," used in the cleansing of mind, body, and spirit. Recent research has provided additional verification of traditional Quinault knowledge, suggesting Oregon grape root can be effective for a wide range of digestive issues, as well as being used for antiseptic purposes or as a topical medication for skin conditions like psoriasis.

The purple-blue berries are tart. They can be eaten one by one or mixed and cooked with other berries. The berries, roots, and other parts of the plant are sometimes used in everyday teas that are both flavorful and medicinal. Wine made from Oregon grape is also reported.

What They Look Like

Tall Oregon grape is a large shrub reaching up to 10 feet high, with shiny, compound leaves with holly-like leaflets, usually 7 to 11 in number, that can turn red in the fall. Dwarf or dull-leaf Oregon grape looks very similar, except that it is low and creeping, seldom taller than knee-high, with 11 to 19 leaflets. Small, bright yellow clustered flowers ripen into bright blue berries. The clusters of tall Oregon grape berries are dense and spreading, while those of low Oregon grape are elongated, on single stalks.

Where to Find Them

Less common within the perennially damp forests near the ocean, Oregon grape is best found in inland mountain forests. Gathering is usually done where there are many plants instead of just a few scattered here and there. When harvesting roots, people often look for well-established patches and plants, gathering near the margins to get the fresh new small roots while not damaging the old "grandmother" plants near the middle of the patch. Tall Oregon grape tends to grow in sunny areas, whereas dull-leaf Oregon grape grows under tree cover.

When to Gather

Oregon grape berry harvesting takes place in the summer, as the berries ripen and become slightly soft. Oregon grape root gathering is most commonly done in the fall or winter, when the plants direct their sugars and active compounds from the stems and foliage to the roots. Justine James Sr. explained that Quinault people traditionally prefer to gather Oregon grape for medicinal purposes in the mountains: "That's where you get it, where it hasn't been contaminated. That's what I do."

Traditional Management and Care

Many harvesters try to selectively gather roots without killing the entire plant, so that Oregon grape populations remain robust. Those who gather Oregon grape for medicinal or other cultural purposes sometimes report showing respects to the plant upon harvesting, or taking steps to ensure the well-being of plants used in this way. Traditional burning, especially on prairies and in interior and mountain settings, likely enhanced the natural abundance and productivity of Oregon grape.

Clusters of lustrous yellow flowers (below), encircled by red stems and prickly green leaves, transform into tasty blue edible "grape" fruits (above) beginning in late spring.

Wild Rose

wleʔnit | *Rosa nutkana* and others

Traditional Uses

Many tribal members enjoy simply seeing the beautiful, delicately scented flowers of the wild Nootka rose in Quinault country—looking like a simple, five-petaled version of the domesticated garden variety of the flower. Yet, the wild rose has many practical uses as well. The fleshy seed pod "hips" are gathered where available. With seeds and inner fibers scraped out, the red-orange outer husk can be eaten like berries. Often tasty and always rich in vitamin C, these rose hips can be eaten by the handful or mixed into pressed berry "cakes" and other recipes. The Nootka rose is by far the most common focus of this traditional use, but other wild and domesticated rose species are also used in these ways.

So too, the wild rose has medicinal value. The stems are used medicinally for sores, being burned down to ash and mixed with skunk oil or other natural oils. The petals and leaves are sometimes used in medicines and teas.

Traditional craftspeople use the pliable wild rose stems for such purposes as babies' cradleboards—medium-sized stems are scraped thorn-free, then bent in arcs that are secured to the board below, or woven together with natural fibers. Tribal members like Lucille Quilt note that the use of the rose for medicine and cradleboards reflects the deep and distinctive cultural significance of the rose to Quinault people—bringing protection and vitality to the child. Many other tribes, in the Northwest and beyond, use rose stems in their cradleboards for similar reasons, sometimes decorating cradleboards with images of roses or other flowers.

What It Looks Like

Nootka rose is an upright branching shrub reaching up to 6 feet. Stems have tiny thorns, and the leaves are oval, coming to a point, with small serrations on the leaf margin. Large, simple flowers have five or more petals in the range of light pink to magenta. Clusters of bright red rosehips mature in the fall.

Where to Find It

Nootka rose is common on brushy roadsides, on the margins of prairies, and on river and streambanks.

When to Gather

Rose hips are gathered especially from late summer through the fall, as blooming comes to an end. This is often the time of stem gathering too—a time when the new shoots from spring have reached their full length and become firm enough to use in crafts.

With their fragrant flowers of light pink to magenta "petals," wild roses are ornamental but also extraordinarily useful in foods, medicines, and traditional crafts.

Traditional Management and Care

Wild rose often responds well to burning or pruning. These activities allow new straight shoots to develop—perfect for craft uses—and can also remove old dead growth and trigger the development of multiple branches, increasing the output of rose flowers and hips. The traditional burning of prairies and other areas in Quinault country likely increased its output. Rose gathered for medicinal or other cultural uses, or for use in cradleboards, is commonly gathered with an appropriate show of respect. Some people look after the rose bushes from which such materials are gathered, and take steps to help the bushes thrive.

Cautions

The seeds inside rose hips are enmeshed in fine hair-like fibers; neither seeds nor fibers should be swallowed, as they can irritate the throat and digestive tract.

Wild rose often grows into entire thickets along streambanks or at the forest edge.

Growing in dangling clusters, bright red rose hips are highly valued as both a nutritious food and in traditional medicine.

Other Important Plants of Meadows, Wetlands, and the Forest Floor

Quinault baskets woven from beargrass, cattail, cedar, and other materials, on display at the Quinault Cultural Center and Museum.

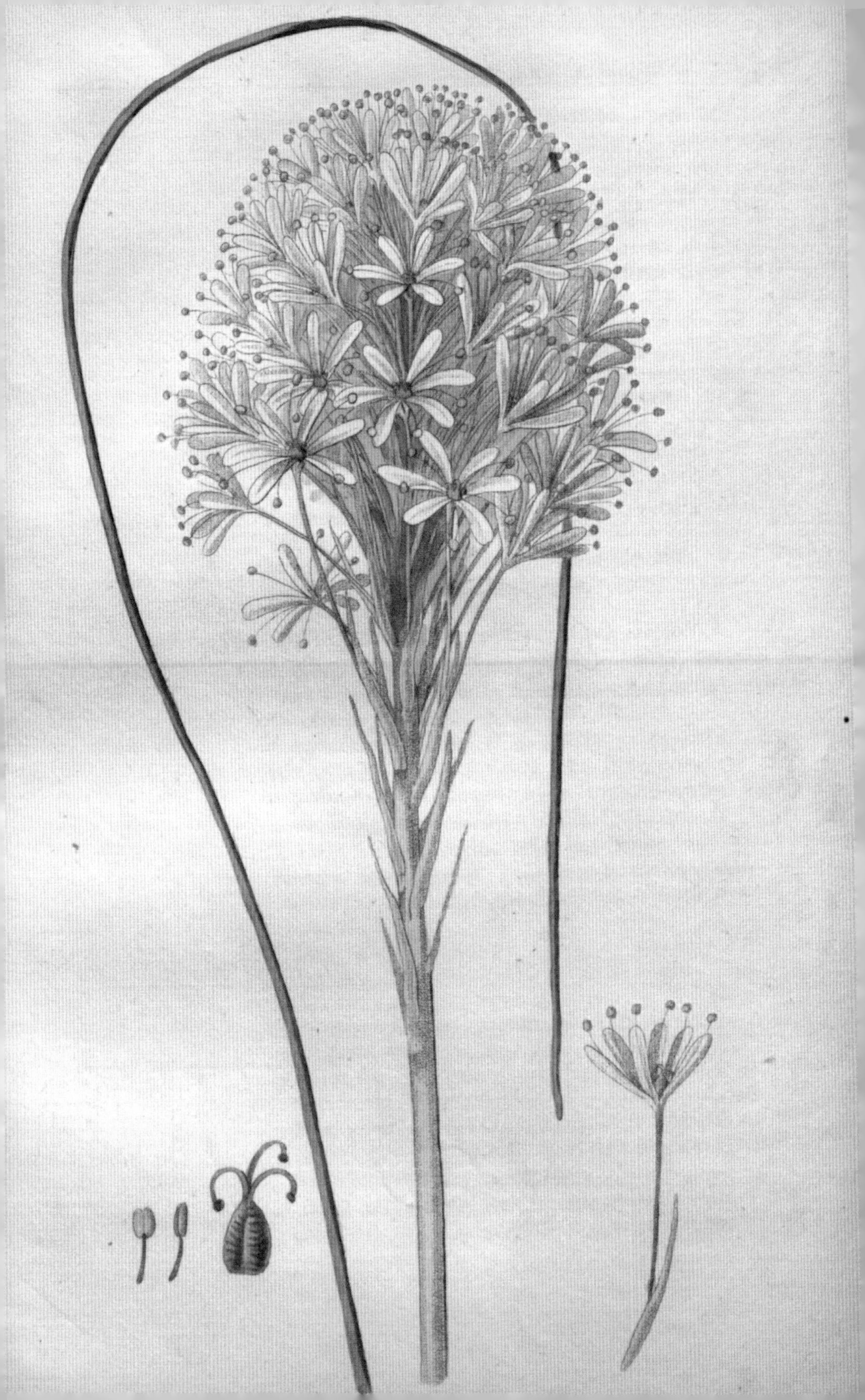

Beargrass

k̓wəlalstap | *Xerophyllum tenax*

Traditional Uses

The tall, white-flowered lily known as beargrass is of vast importance in Quinault history and in cultural traditions. The long, tough leaves are used in basketry, especially serving as the white weave or decoration on baskets, and are used in many ornamental objects and regalia. Harvested widely in subalpine settings throughout the Pacific Northwest, beargrass is called by many names, including "mountain grass" reflecting its usual location, or "cut grass" reflecting its sharp leaves; in Canada, it is often called "American grass," as it is relatively uncommon north of the border and is traditionally obtained from American tribes. Long ago, glaciers pushed entire intact patches of beargrass down the Olympic Mountains, depositing them practically in the Quinaults' backyard. Using and caring for these patches over generations, Quinault families had access to conveniently large and low-elevation patches of beargrass, a situation not found in other tribal territories within the region.

Harvesters traditionally seek the tallest plants with the longest leaves and, typically, plants that are not flowering. The leaf blades in the middle of each clump are the most pliable and are therefore sought traditionally by weavers, with perhaps only two or three of the innermost leaves being suitable for harvest. The strong, serrated leaves are extremely sharp and must be pulled with care. Knowledgeable harvesters reach into the middle of the clump carefully, wrapping the flexible inner leaves one turn around their (usually gloved) hand, and pulling sharply, causing the leaves to separate from the rootstock at their base. These leaf blades are then tied together in small bundles. At home, these are untied, re-sorted according to size, and rebundled; the bundles are then hung out to dry, usually in direct sun to bleach the strands a brilliant white. Harvesters allow the bundles to dry over the course of several days, some bringing the beargrass

Beargrass thrives in the subalpine meadows of the Northwest, though Quinault country is uniquely blessed with low-elevation patches, sustained by burning and other forms of traditional care.

"I used to go pick beargrass with the elders. . . . You just reach up there, grab the center, the heart of it and pull that out and leave the rest. . . . They never killed the plant. They never killed the plant on anything."—Gerald Ellis

"In summer the Quinault made ocean voyages to the Chinook country, stopping the first night out at Tshels (Point Hanson), and the second at Cape Shoalwater, and on the following evening rounding into the estuary of the Columbia. The Clatsop village Nu'smá'spu, on Youngs bay south of Astoria, was a favorite rendezvous for visitors from the north, on account of its nearness to the trading-post [at Astoria]. Here the Quinault gave Makah canoes, slaves, dentalium shells, sea-otter skins, beaver-skins, otter-skins, baskets, and a variety of coarse grass (probably bear-grass), in exchange for the goods of the white traders."
—Edward S. Curtis, 1913 (based on interviews with Quinault elders)

indoors at night to speed drying. Once all leaf blades are dry, they are stored for later use. Before being woven into baskets, the hard middle strand of the leaf is removed by weavers who scrape and trim the blades to the size needed for the basket. While bright white is most common, some tribal members boil beargrass with natural or artificial dyes to produce basket decorations of various colors.

Traditionally, bundles of beargrass are popular gift and potlatch items. Many weavers still exchange bundles today. Beargrass was a tremendously important trade good historically, contributing to the wealth and status of Quinault people, and generations ago, Quinault families paddled to trading centers on the lower Columbia River with canoes filled with beargrass. They also traveled north, bringing beargrass to trade with the Makah, Nuu-chah-nulth, and other people to the north. During the international fur trade, beargrass could be traded with the Chinook and other Columbia River peoples for imported goods such as guns, beads, and blankets—one of the earliest ways Quinault families obtained these items. In lean times, beargrass could even serve as a food, as its roots are edible to humans, as they are to other creatures.

When growing in the wild, beargrass responds well to fires, which leave the roots intact but help remove competing vegetation, adding nutrients to the soil and improving the germination of the plant's seeds. Fire also renders the leaves more pliable and thus more ideal for basketmaking. During a fire, the outer leaves of a beargrass clump may become crispy while those in the middle are protected—another reason the middle strands are preferred. For at least a thousand years and probably longer, Quinault people traditionally burned patches of this plant to maintain beargrass, show respect, and increase the plant's output. This is a long-standing practice. In recent years, the tribe has returned to active management, burning some patches and monitoring the results.

The Quinault Indian Nation has worked hard to stop illegal and disrespectful harvests of beargrass. Illegal harvesters sometimes take truckloads of beargrass to buyers who, in turn, ship the foliage oversees for use by florists. Though prohibited by law, this harvest has greatly harmed some patches, especially along Highway 101. Many non-Indian harvesters not only take too much, but also sometimes remove all of a plant's leaves, even pulling plants up by the roots, killing them outright. Today, some basketmakers obtain their beargrass not through traditional harvesting but by receiving confiscated materials from the Quinault Indian Nation.

What It Looks Like

Beargrass has tough, toothed, grass-like leaves up to 4.5 feet tall, but more commonly shorter, with sharp edges emerging from a central clump. A cone of numerous tiny white flowers stands aloft on a stalk reaching 2.5 to 6 feet, with clusters of plants usually blooming together every few years at the most.

Where to Find It

Beargrass can be found in subalpine meadows throughout the American Pacific Northwest. Prairies along the Olympic Peninsula coast were once productive gathering areas. Occasionally, beargrass can be found in the understory of forests from sea level to above 6,500 feet.

When to Gather

Beargrass is especially gathered in the summer and fall, when the central new leaf blades have reached full length but are still pliable and strong. Peak gathering is said to coincide with elk hunting season.

Traditional Management and Care

To reach their full historical scale and productivity, beargrass patches required regular burning. The patches visible today are only a fragment of Quinault's once rich beargrass prairies. Traditional harvesters gather only the new central leaves and keep the rest of the plant in place, allowing the plant to survive, thrive, and grow new leaves into the future.

"When you look . . . for beargrass there's not a lot of patches out there, not as much as you'd like. It's mostly for our weavers, who make baskets and things like that. I don't think there's that much out there [nowadays]."—Marlene Hanson

Cautions

Beargrass has very sharp edges, so harvesters commonly use gloves or other hand protection when gathering, processing, or using this plant. Some harvesters note that they gather during the hunting season, and in places that are actively hunted, such as prairie clearings in the forest, so they must be careful to wear bright clothing.

Sweetgrass

kaktsiw | *Schoenoplectus pungens*

Traditional Uses

Sweetgrass is one of the most important basketry plants used by Quinault people, both historically and today. Stronger and more flexible than common basketry materials like cattail, sweetgrass is used in many different ways. The sedge, resembling a grass, often serves as the foundation and weave of baskets, but can also be used as filler in coil baskets made from other plant materials. Additionally, sweetgrass is woven into mats and other durable woven goods for household purposes.

Harvesting has long centered on certain estuaries known for the quality and abundance of sweetgrass. In especially large, enduring bayshore patches, a diversity of plant sizes and textures predominates. The sweetgrass fields are vast enough to have endured and supported intensive harvests over centuries. Historically, Quinault families visited the shoreline every year to harvest—revisiting places used and managed by families over the course of generations and bringing home canoe-loads of sweetgrass. Quinault families maintained seasonal camps there, drawing fresh water from streams that flowed into the bay. Though access has declined along the entire bay, certain harvesting areas continue to be popular sites for gathering, revisited by many Quinault families today. At these sites, sweetgrass is picked at low tide, when patches are exposed or only slightly submerged. Intrepid harvesters even access hard-to-reach patches in deeper spots by extending a board into the water to be used as a foot plank. These deeper plants are coarser, harvesters report, and often used for thickly woven baskets and mats or for other purposes demanding thicker materials.

The harvesting of sweetgrass has its own unique methods and etiquette. Harvesters look especially for long sweetgrass leaves, but also gather shorter leaves for specific purposes. They do not take the flowering stalks and do not take all of the leaves—allowing plants to survive and thrive in spite of the harvest. Leaves of sweetgrass are pulled by hand and never cut with a knife. If the plant is cut, an important part of the stem or leaf's outer sheath is left behind, meaning it will not dry properly and will be unusable. Thus, experienced harvesters recommend pulling the stems gently and slowly while holding the plant close to the root. Families maintain regular paths through the best patches, and people are encouraged to stay on those paths as much as possible to avoid trampling. Children who can't stick to the paths are encouraged to stay out of sweetgrass patches entirely. The amount of sweetgrass gathered by each harvester depends on weavers' needs and storage space, with many families gathering enough for a year or more in a single visit to a favorite patch.

"The sweetgrass . . . I learned under Lillian Pullen, but I just now started harvesting it. You harvest it three different times . . . the beginning of July, the end of July and then in September. That way, each of the colors will be different as the seasons change. It turns that golden color. You never cut it. You pull it. You stay on the trails that other weavers have been on. It's like hay or wheat. If you trample it, you're bruising it. Children are to remain up on the shore. They are not allowed to go down with the basket maker. You always pull it—you never cut it. And you don't be too greedy. One time I got too much and it got moldy. . . . Sometimes I felt like 'Oh, I got sabotaged for making that stuff go moldy!'"—Harvest Moon

The processing of sweetgrass also requires special care. Once harvested, the stems are washed carefully, first at the shoreline and then more intensively at home. Often modern harvesters use soap and water for this purpose. Usually harvesters remove the natural outer sheath from the base of the sweetgrass, as it will stain the stems a reddish color on one end. Harvesters then spread out the plant materials to dry, for if the sweetgrass is not completely cleaned and dried it will mildew or mold. Near the moist ocean coast of Quinault country, drying is done mostly indoors. Tribal members in drier inland places can sometimes dry their grass in outdoor shade if weather permits; though if sweetgrass is dried in the direct sun, it will often become too brittle for use. When the stems are completely dry, weavers tie them in bundles, often hanging the bundles where they will continue to stay dry.

Before using sweetgrass in a basket, weavers scrape the strands to remove dirt or the hard outer coating of the stems, making them more flexible. When preparing to weave a basket, weavers can also soak the stems in water to make them more pliable and to reduce their sharpness. For visible places on a basket, weavers use only clear light-colored strands, avoiding those with imperfections or dark spots that tend to be brittle or that give way to mold. Still, moldy strands can sometimes be washed. If they come clean, they can often be successfully redried and reused. If they remain discolored but mold-free, they might be salvaged with two strands wrapped together to use as the thick and mostly unseen "warp" of the basket. Sweetgrass does not take dye as well as some basket materials, but naturally ranges from light yellow to green to brown, depending on how it is dried. Faster drying results in brighter, lighter colors.

"You bring [sweetgrass] home and rinse it off. Some people soap it down and then rinse it off. And then you have to lay it out, like in your yard, just, like, for the rest of the day, maybe the next morning. And you have to get it out of the sun and then lay it around in your house everywhere you have space. It only takes a few days to actually dry. And I just put it in hand-sized bundles and tie it. Because when you're gathering it too, you just take and you rip material and carry it out that way, and then tie the bunches together. And you can hang it or just leave it laying. Prop it up against a stump."
—Alicia Figg

In more recent times, people have increasingly picked an entirely different "sweetgrass" (*Hierochloe odorata*). This is a finer-grained true grass, common in both fresh- and saltwater wetlands in Quinault country and many parts of North America. This "sweetgrass" has grown in popularity in part because tribes throughout North America use the plant, which produces attractive light-colored baskets. Many of the techniques used to gather and process the more common *Schoenoplectus* sweetgrass are also applied to *Hierochloe*. Leaves are taken instead of stems. Dried with the same caution as the other sweetgrass, the leaves dry from yellow to green depending on when the grass is picked and how quickly it is dried. With its light, fine-grained look, it has become popular in decorative baskets, as well as in the production of dolls and other ornamental goods. Experienced weavers often mix sweetgrass dried in different ways to create subtly colored designs, or add a few dyed fibers to make especially rich decorations. Elders note that it is especially important for harvesters to pick *Hierochloe* sweetgrass with caution, as the leaf blades can cut a person's fingers if harvested quickly and recklessly.

Both types of sweetgrass, and the baskets they form, have always been important trade items. Even long ago, bundles of these plants, especially *Schoenoplectus*, were used in trade with tribes such as the Makah, who lacked easy access to large patches of the plant. In the nineteenth and twentieth centuries, some Quinault families made extra income by selling sweetgrass baskets of both species to non-Indian visitors. As with other uniquely productive plant harvest areas, sweetgrass gathering areas on large bays in Quinault country are traditionally shared among tribes. Even in recent times, Quinault tribal members harvest there alongside families from Quileute, Makah, Hoh, Skokomish, Squaxin Island, Chehalis, Chinook, S'Klallam,

"I learned quite a bit from my aunt, my great-aunt and my great-great-aunt, Louisa Polsper. . . . She taught us how to pull sweetgrass and all that, without getting cut. She'd say, 'Come on, I'll show you.' And she took me out there and she went down really slow, and pinched it, pulled it up. [And again] went down really slow, pinched it, pulled it up. Then I start going faster, that's when I start getting cut. She'd say, 'Slow down, slow down: it's not going to go anywhere!' . . . My great-great aunt told us not to cut them, just pull them straight out."—Dennis Martin

and other tribes. Tribes and families unable to travel to those uniquely productive sites instead trade for weaving materials gathered from such places.

What They Look Like

The traditional Quinault sweetgrass (*Schoenoplectus pungens*) has upright pointed dark-green stems, thick and sharply triangular in cross section, each connected by creeping underground rhizomes. Oblong, pointed, beak-like flowerets poke out from single small clusters on the end of the fruiting stalks. The other "sweetgrass" (*Hierochloe odorata*) is an often fragrant perennial grass with solitary stems that pop up from creeping horizontal rhizomes. The flat, soft leaves can reach up to 30 inches tall; flowering and fruiting heads are borne at the top of a long central stem.

Where to Find Them

Traditional Quinault *Schoenoplectus* sweetgrass grows best on wide grassy tide flats of low slope, where freshwater streams help moderate salinity. The most important patches are found on riverine estuaries. Harvesters note that the extent of sweetgrass patches on the bays has changed over the years because of contamination, changes in runoff, and other effects from development. Some advocate restoration, or new plantings in suitable habitats, to avoid these effects as well as undue pressure on areas used by tribal members. The other "sweetgrass," *Hierochloe*, is generally less concentrated in any one location, being found on marsh edges but also on the edges of lakes, bogs, streams, and other settings within Western Washington.

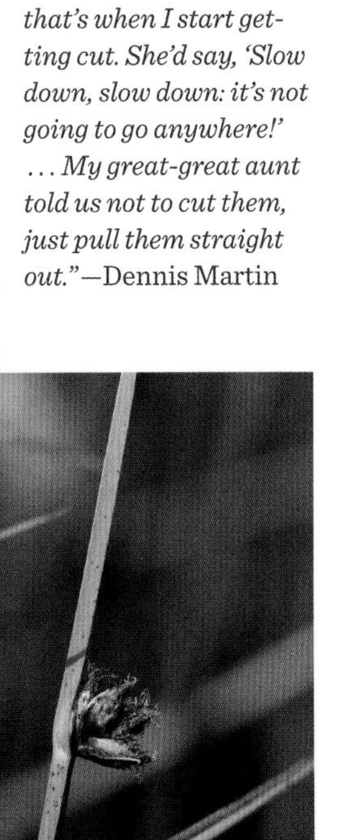

Schoenoplectus, the sweetgrass most commonly and traditionally used in Quinault basketry.

When to Gather

Harvesters report that they gather both types of sweetgrass in the summer, especially in July and August, but sometimes as early as June or as late as September. *Schoenoplectus* sweetgrass harvested at different times of the summer often dries differently and therefore has subtly different colors. Drying can be difficult with materials harvested very early or very late in the season, causing the sweetgrass stems to become mildewed or brittle.

Traditional Management and Care

Harvesters say that sweetgrass can be harvested in large quantities if this is done respectfully and selectively. Staying on trails and not trampling the sweetgrass is important, as is harvesting and processing sweetgrass appropriately so that none is wasted. Harvesters mention that "word gets around" if people are improperly harvesting and wasting sweetgrass, or mismanaging patches.

Cautions

Growing in the salt marshes, where many things settle out of the water, sweetgrass is said by elders to be more threatened than many culturally important species by pollution, pesticides like Sevin (carbaryl), the invasion of introduced Spartina grasses, and a range of other outside effects. In some settings, elders express concern that pollution may be severe enough that it could affect harvesters; careful washing of materials may be advised. Sweetgrass often grows near places with deep mud, where people can get stuck in the muck.

Hierochloe sweetgrass, an important plant in the basketry and ceremonial traditions of Quinault and other native nations across northern North America.

Harvesting sweetgrass in the estuarine margins, tribal youth learn both the mechanics and ethics of traditional harvesting from their elders.

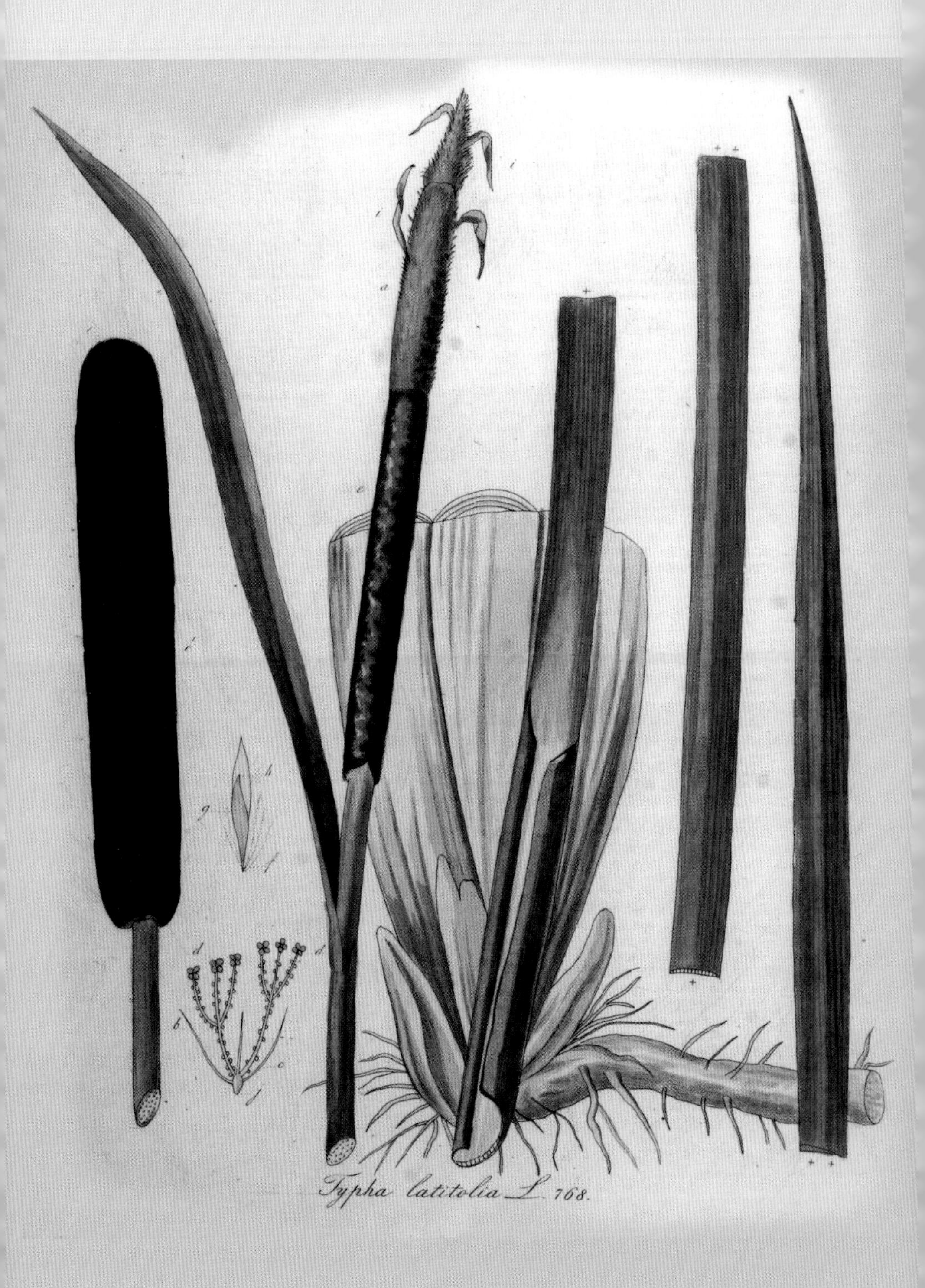

Typha latifolia L. 768.

Cattail

swiči | *Typha latifolia*

Traditional Uses

Cattail is among the most culturally important plants to be found, and a keystone of Quinault traditional crafts. The long, flat leaves of cattail are woven into large mats that are historically of vast importance: as roofs and walls for temporary structures to be raised when guests arrive; rolled like tents to be carried from place to place by canoe; as "carpets" spread on the longhouse floor; as mat partitions within longhouses, separating spaces into rooms; as a liner for drying or cooking food; or as a placemat-like surface for eating. When traveling, people use cattail mats as kneeling pads for canoe journeys, or folded like gunny sacks for packing light items. Additionally, the newest, thickest cattail mats are traditionally used as mattresses. In cold weather, such mats sometimes provided insulation against exterior walls of a longhouse or were hung over doorways to control drafts. At one time, almost every Quinault home had several cattail mats.

Woven cattail was used for many other purposes. People formerly braided the leaves in a manner similar to cedar bark: making conical waterproof cattail rain capes and coats, corded cattail skirts for women, cattail hats and headbands, and sometimes even woven cattail ceremonial regalia. Pieces of cattail, split long and thin from the edges of the lower part of the leaves especially, could be rolled against the thigh, making a durable cord that was used in fishing lines, in baskets, or for sewing.

Among the traditional uses of cattail, though, perhaps the most visible and enduring is in basketry. Many cattail baskets are made with a loose weave, producing a lightweight and light-duty basket. Cattail could also be woven in a simple open weave for such uses as light-duty food baskets, allowing water to pass freely through the open weave to wash or drain the basket's contents. Tighter weaves have been used for other basket types, with cattail often serving as a "filler," closing the gaps in baskets between stronger, thinner woven materials such as spruce root. In baskets made of the other materials, cattail has been used as the "foundation coil" at the base of the basket. Many cattail baskets have a

"Oh, I just watched my mother and grandma and they taught us how to make the baskets [with both cattail and sweetgrass]. Yeah we learned from [the time we were] small. We learned how. We were alright. That's the way they taught us how to make the baskets. I was about seven or eight years old . . . when we had to learn how to make a basket."—Maggie Kelly

characteristic thick, ropey, light-green-to-beige look, in part because women weave cattails in their natural color. Not only are the plants difficult to dye, but they look appealing in their natural state. The use of cattail for baskets remains a lively and thriving cultural practice in Quinault country. Women still teach children how to make baskets at an early age, also teaching methods of gathering and processing cattail from areas of ancient importance.

When gathering cattail, people generally look for the longest leaves available, cutting the leaves (not the flower-bearing stems) as close to the base as possible. Women often pick large quantities of cattail in a single trip—sometimes a pickup-truck load or more. Before automobiles were available, women once packed large quantities of cattails back from the marshes on their backs. If they were fortunate, they had access to productive cattail patches within an easy canoe trip of home. Those who were not physically able to make these journeys recruited younger gatherers to assist. Elders such as Clifford "Soup" Corwin recalled that elders of their youth often gave baskets or other woven items to people in exchange for freshly harvested cattail.

Cattail, commonplace in Northwestern wetlands, is a cornerstone of traditional Quinault weaving, used in mats, capes and coats, baskets, and many other traditional goods.

After taking cattail home, harvesters dry the leaves and use the stockpile for weaving baskets and other items through the year. Harvesters immediately clean and process their cattail leaves, promptly peeling the exterior to remove the fleshy, slimy insides. The cattail leaves are then dried indoors, or outdoors in the shade on warm, sunny days. Left in the sun for long they become brittle, and left out in damp weather they quickly mold. After a week or so, when the cattails are dry enough to be storable and bendable without breaking, they are bundled indoors for later use. Dried cattail is, and has always been, a trade item—often exchanged for unique basketry materials found in other tribes' territories.

Beyond this, cattail is edible—the white, soft bases of the new spring and summer shoots can be pulled up and eaten. These have a nutty taste, vaguely potato- or cucumber-like, and the shoots are traditionally dipped in grease or combined with other foods. The roots are also edible, being cooked, peeled and eaten, or ground into a flour-like powder to use as a thickener.

What It Looks Like

Cattail is a 4-to-6-foot-tall herbaceous perennial that grows in water, with tall, flat upright leaves growing in tight clusters and stems carrying a brown rodlike spike of female flowers—the "cat's tail"—topped with a thinner spike of male pollen-bearing flowers. With maturity, the brown spike breaks apart to seedy fluff.

Where to Find It

Cattail is very widespread at low elevations, sitting in slow water such as marshes, swamps, seeps, and ditches and on the edges of ponds, rivers, and streams. Many people have their favorite places to gather. The Quinault practice of harvesting cattail in bulk on the shore of Grays Harbor and other bays predates reservation times.

When to Gather

People start gathering cattails when the leaves have reached their full size in early summer and continue until the plants stop developing full, new leaves—late June through August is peak season. Later harvesting is possible where the leaves are slow to develop, such as in shady areas. Leaves gathered too early will shrivel when drying, while those gathered too late often become brittle.

Traditional Management and Care

People often return to the same patches of cattail year after year, developing long-term connections and understandings with these particular plants. Women are careful not to overharvest particular patches and often gather only selectively from individual plants so that harvesting will not kill the plant. Filling, spraying, and other wetland interventions can deplete or eliminate cattail patches of long-standing importance to Quinault families, prompting some modern harvesters to get actively, even politically, involved in the protection of these sites.

Cautions

Some weavers advise against gathering cattail in ditches near roads, or in bays near cities or other developed areas, as these plants are potentially contaminated with polluted runoff. Tribal members share similar concerns about gathering plants in such places for food use, noting that cattail absorbs pollutants in its roots and shoots. Permits are required to harvest in certain places.

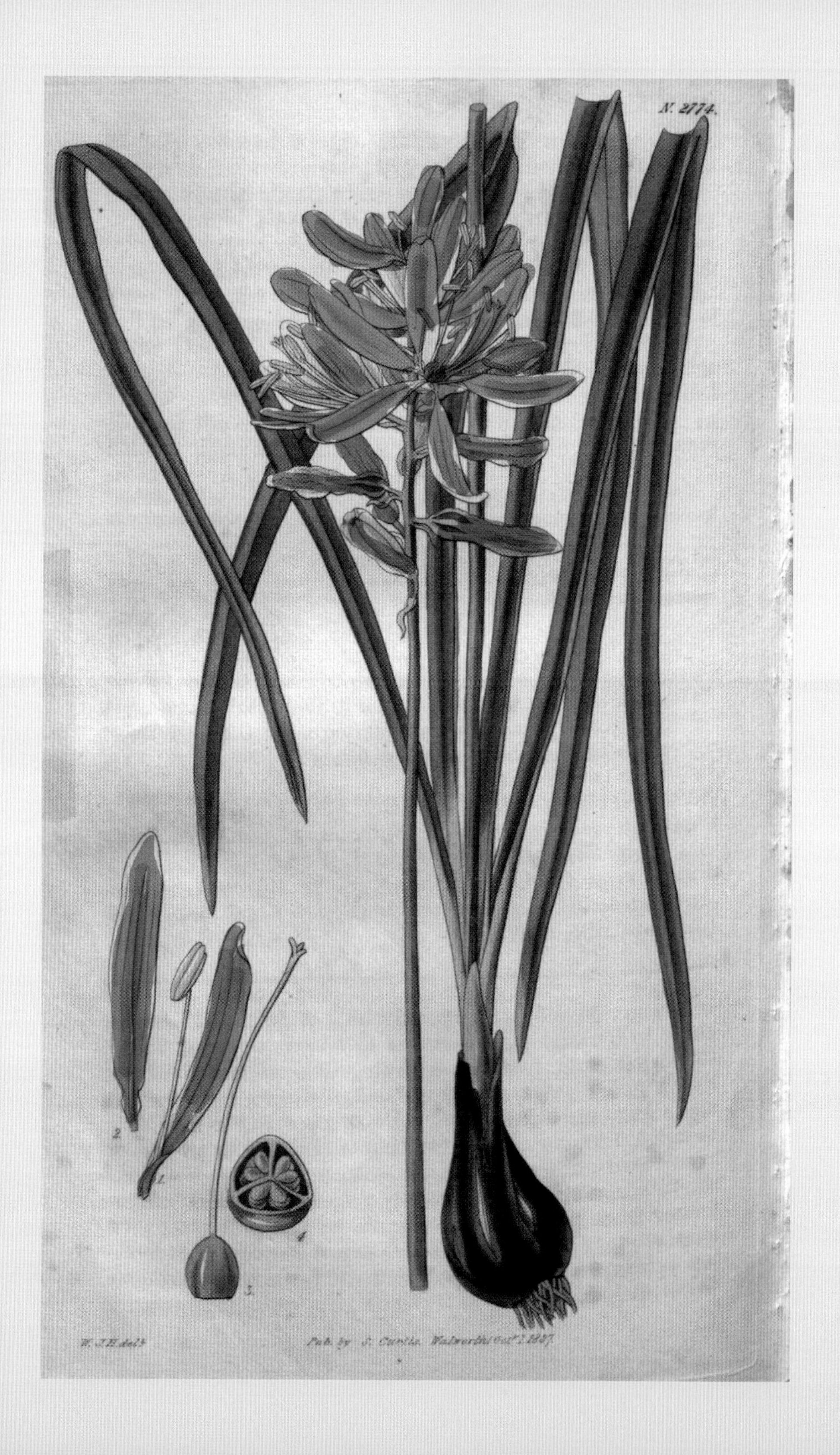

N. 2774.
2
1
4
3
W. J. H. delt
Pub. by S. Curtis. Walworth Oct. 1 1827.

Camas

milaakels

C. leichtlinii (great camas)
Camassia quamash (common camas)

Traditional Uses

Camas is one of the most delicious and culturally important traditional food plants of the Northwest. Dug up and eaten raw, camas bulbs are bitter, sticky, and not especially digestible. When roasted, however, the bulbs turn soft and dark, their sugars caramelizing to produce a range of rich flavors that some people compare to cooked pear, pumpkin, figs, or sweet potato. It is easy to understand why camas was once a staple food of Quinault and other Northwestern tribes and a cornerstone meal of traditional feasts and potlatches.

Camas is traditionally roasted in pits with hot stones, sometimes doused in water to produce extra steam. The bulbs were wrapped or piled between sword fern and salal leaves to impart a pleasant flavor. The cooked bulbs can be eaten one by one, though historically Quinault families mashed them, then pressed the mash into loaves that later could be cut and eaten like a pudding. Often these loaves were recooked and partially dried before storage, then reheated before eating. Camas bulbs and camas loaf "puddings" were dipped in oil traditionally. In recent times, Quinault families have developed new recipes for cooked camas—making camas pies, camas soups, or preparing the mashed bulbs as they might prepare a sweet potato casserole.

Traditionally, camas was dug from the ground with specialized digging sticks of yew, spruce, cedar, or other woods, though shovels are common today. Harvested bulbs are placed in open-weave baskets and washed in water to remove the soil. They can then be cooked immediately or stored for later use. To be preserved, they can be dried raw, though people often process and cook the bulbs before storage, so that camas can be quickly reheated when needed. The outer skin of the bulb is peeled before

"The roots were stored in baskets and when a good supply had been secured, the family returned home to the village. Here the roots were washed by placing them in an open basket and running water over them. A pit was then dug in the sand, partly filled with rocks, and a brisk fire built on top. A large supply of fresh fern leaves was gathered and when the rocks were thoroughly hot, the fire was removed, the stones leveled, and a layer of fern leaves spread over them. On top of this bed the camas was spread (the size of the pit depending upon the amount of the roots). A thick layer of fern leaves was placed on top and covered with a three-inch layer of sand. The roots were allowed to cook overnight and a fire was kept burning on top of the covering sand during the baking. In the morning the roots were removed, mashed, and made into cakes about twice as large as a loaf of bread. These cakes were buried in the reheated pit between layers of fern leaves and baked for a day. They were now thoroughly cooked and would keep through the following winter. The cakes were wrapped in a grass called pȧla'pȧla [probably sword fern] and stored away. Slices were cut off as desired. The Quinault were extremely fond of camas and regarded it almost as a confection. The feeble and ailing regarded the freshly baked roots as very beneficial, and special trips to the 'prairies' were often made to satisfy their whims."
—Ronald Olson, from Quinault elders' accounts, 1936

or after cooking. Once cooked, individual bulbs are often dried. Historically, families kept dried bulbs in boxes or large baskets, pulling them out to soak in warm water for a few hours, softening them before eating. More recently, the bulbs are stored in large gunny sacks. Families sometimes keep large bags or baskets beside the fire pit or wood stove in their homes so the camas will be warm and ready to eat throughout the day.

In Quinault country, camas is traditionally harvested in fire-managed prairies and other meadow clearings. The camas prairies are important cultural sites, appearing often in ancient oral traditions. For example, Quinault story cycles describe Eagle proposing that a huge camas prairie be placed where Lake Quinault is today, though Raven objects, saying, "No, that would be too easy for the people; they ought to work if they want anything. If they want camass-root, they should be compelled to go through the woods and find the prairies, and pack the camass out."

Indeed, the Quinault people did work to keep their camas prairies productive: camas was once managed and harvested intensively in the small prairies of the Quinault coast. From these clearings, Quinault harvesters have selectively taken only a portion of the bulbs to ensure strong future harvests. They have used fire to remove competing vegetation and release soil nutrients; they have reseeded and even transplanted plants. And they have intentionally churned the soil, as looser soil is both easier to dig and encourages larger bulbs. Quinault families once relocated for days or weeks to these specially managed camas patches, digging and processing bulbs together while living in temporary shelters. They camped not only beside other Quinault people, but often alongside people from other tribes who visited and were invited to partake in the harvest at the most productive camas prairies.

The common and scientific names of this plant

both entered worldwide usage through Chinook Jargon (*Chinuk wawa*), which seems to have adopted the term from the Nuu-chah-nulth words, *cha-mas* (fruit) and *cha-mas-sish* (sweet). Camas was a trade good of huge importance and once was among the few foods to be traded almost constantly between tribes—second only to high-quality pieces of dried salmon in trade importance. With Quinault's productive prairies, the tribe was said to have once held a near-monopoly on Washington's ocean coast camas trade. Quinaults gave raw or cooked bulbs to camas-poor neighbors such as Makah in exchange for a variety of trade goods. And when Spanish and other early European ships passed through the area, Quinault people presented them with camas to trade. In the 1800s, some Northwest tribes resisted, or even fought wars, when camas patches were taken over and destroyed by arriving American settlers. In recent years, the plant has not only regained popularity as a culturally important food among Northwest tribes, but is increasingly sold as an ornamental plant for gardeners in the Pacific Northwest and beyond.

"In the prairies . . . most of the time they [burned] for the replenishment for the deer and elk and bear . . . and that root . . . camas . . . it comes back better. All the nutrients just fall to the ground and everything sprouts up, all the seeds that were there. Some of them need to be burned in order for them to sprout." —Dennis Martin

What It Looks Like

Bunches of thick, grass-like leaves emerge from a single base at the top of the camas bulb. Showy, bright blue flowers are arranged loosely atop a single stem reaching 1–3 feet high, with common camas on the shorter end of this range and great camas on the taller. The flowers of common camas are slightly asymmetrical, while great camas flowers are larger with evenly spaced petals. The edible bulb of both species, onion-like in appearance but not in flavor, sits just below the ground, but up to 6 inches deep in deeper soils.

Where to Find It

Camas grows seasonally in moist open meadows, prairies, and grassy slopes from low to middle elevations. The plant lives largely in damp areas that ancestors burned historically, including lowland prairies and meadow clearings in the Olympic Mountains; many modern plants descend from patches tended by these ancestors and, in some places, were likely transplanted to those locations.

When to Gather

Camas is often gathered when the plant has most of its sugars and nutrients concentrated in the bulb, especially during the spring (March-April) and fall (September-October). Yet elders mention other harvest times, such as in June, when Quinault, Queets, and Quileute people traditionally gathered together on certain prairies to dig camas. Elders such as Katherine Barr say that camas is ready to harvest by summer—"when [the camas] start dropping their seeds, that's when they're ready"—but that the best time to seek camas is in the late summer and fall, when most of the berries "are starting to dry up" on the bushes. Bulbs gathered in the fall dry better than those harvested in the spring and can be stored for long periods. As camas leaves and blooms are largely gone by fall, people sometimes mark the plants they want to dig with sticks or other markers so the bulb can be found; dry seed stalks also commonly remain to mark bulb locations.

Traditional Management and Care

Few plants in Quinault country have been the focus of so much careful traditional management as camas, which

Small camas bulbs, freshly dug from a traditionally managed prairie in Quinault country.

The striking flowers of camas appear on stalks each spring, briefly turning entire fields purple-blue.

was actively cultivated long before European people arrived on this coast. Quinault people have burned prairies and other clearings in part to keep the forest canopy at bay and allow the camas to thrive. When harvesting, people are selective, taking only a few plants from each bunch and leaving many behind so as to not depopulate a patch. Harvesting when the seed pods are still standing in summer and fall can help disperse seeds and maintain good patches, and some families have traditions of scattering the seeds as they dig. In some cases, families appear to have transplanted bulbs or seeds from place to place in order to bring productive camas patches close to home. People also intentionally churned up the soil at these places, knowing that this would make future digging easier and allow larger bulbs to grow. Elders also traditionally weeded out "death camas," which often grows in camas patches (see "Cautions," below). Receiving so much care, camas is a traditionally cultivated plant: left alone, it does not usually thrive in Quinault country and will be overgrown by forest or shrubs. It has also been overgrazed by livestock or otherwise harmed in portions of Quinault country. As Jared Eison notes, "It doesn't take much to wipe it out."

Cautions

Beware: the creamy-white-flowered "death camas" (*Zigadenus venenosus*) has bulbs that look very similar to camas and grows in similar places, sometimes even in the middle of edible camas patches. Death camas is highly poisonous, though much less common than camas. As Conrad Williams puts it, "If you pick the wrong one, don't worry about dessert after dinner, because you won't be around!" Death camas can be distinguished by its creamy-white flowers, usually growing more densely than blue camas, and with shorter flower petals than the edible camas. If in doubt, check online photos or guidebooks; if doubts remain, avoid suspicious plants completely. Traditionally, elders weed this plant out of camas patches, gradually trying to eliminate the hazard it presents. Without their labors, it is possible death camas would be much more common and problematic today.

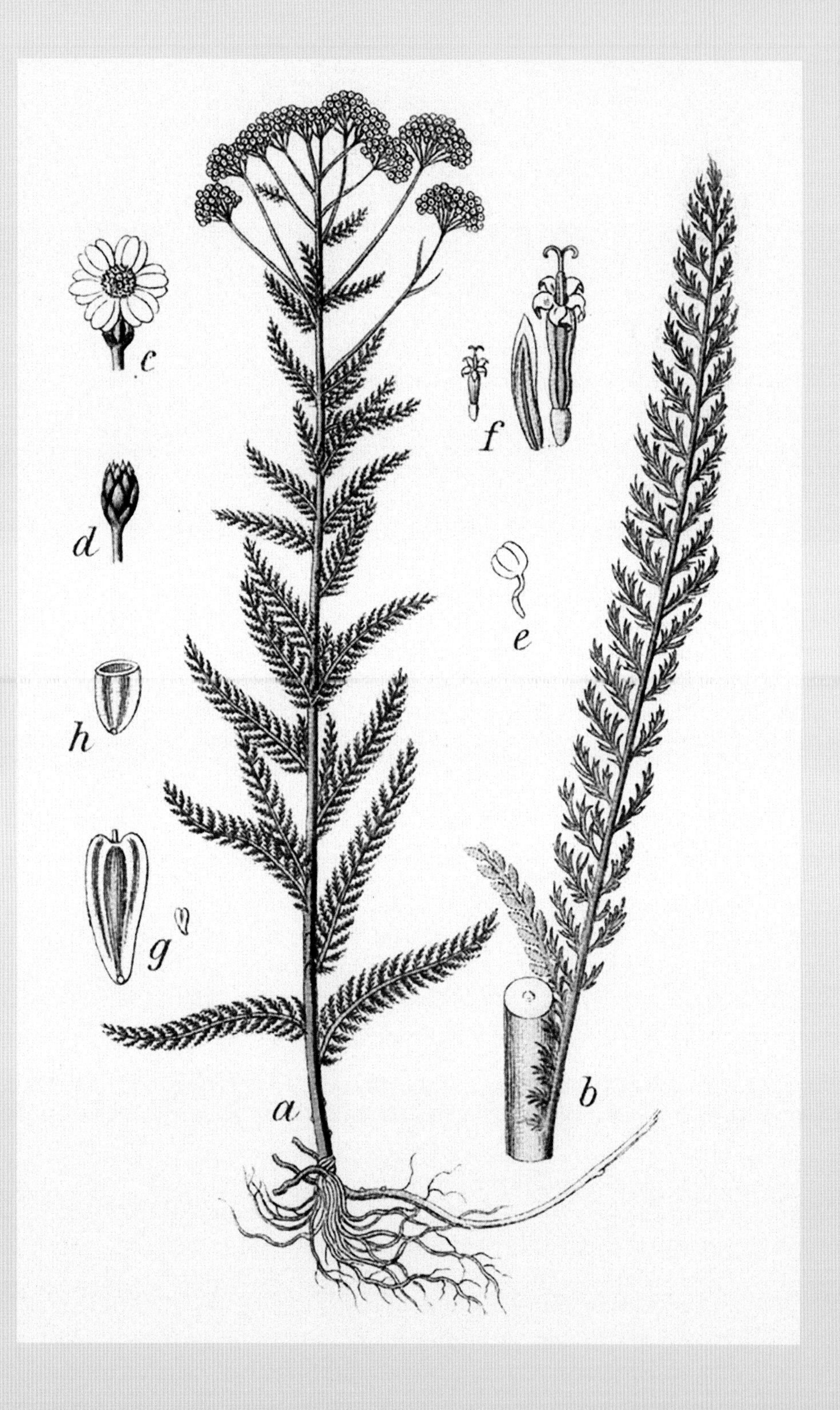
c
f
d
e
h
g
a
b

Yarrow

leko?stap | *Achillea millefolium*

Traditional Uses

Yarrow, with its frilly leaves looking like little green squirrel tails, is among the most popular medicinal plants for the Quinault. The plant is used topically, ingested, and even inhaled, allowing Quinault plant users to benefit from its celebrated antiseptic and anti-inflammatory properties. This plant is found in many places around the globe, and is among the most widely used medicinal plants worldwide.

Many use this plant as a poultice for healing wounds and bruises. For this reason, the plant is often chewed or ground up to release its active compounds. Elders report the plant has powerful antiseptic properties, applied to cuts and other wounds to help prevent infection. Moreover, it is often bound to the wound until it has healed. Elders also report that the plant, if kept bound to a wound, will help reduce scarring. When used topically, yarrow has anti-inflammatory potentials that are widely known; heated slightly, a poultice of yarrow leaves is applied to joints swollen by injuries or by arthritis. When the poultice is especially hot, or moistened and steaming, it can be wrapped in thin cloth to protect the skin. Quinaults have a tradition of using such poultices for animals too, applying yarrow to the strained joints and muscles of horses or dogs. A warm, moist yarrow poultice is sometimes applied to reduce fever in children and adults.

Yarrow is also widely used in teas, tinctures, and other decoctions. Slow-simmered yarrow roots and foliage are widely reported to produce a healthful tonic tea, good for reducing bodily inflammation and for soothing sore throats and other cold symptoms. At times the plant is also mixed with other ingredients, such as bitter cherry bark, to aid in fighting cold symptoms. Mixed with dandelion and possibly other species, strong yarrow teas are even said to fight diabetes. A liquid decoction from this plant is reported as a good source of soothing eyewash, as well as a medicine consumed for tuberculosis and other respiratory ailments. And a boiled decoction of yarrow has been used to help smooth and speed childbirth, especially when a woman has been in protracted labor (but see "Cautions" below). Women sometimes chew yarrow for the same reason, causing sweating and hastening the birth. After childbirth, women sometimes continue to chew yarrow to aid in healing the uterus and cleansing the body.

In each case, the strength of the infusion is chosen to match the malady. In mild decoctions, yarrow is often mixed with other herbs to make a pleasant and healthful tea. At medium strength, yarrow tea is often reported to aid in diarrhea and upset stomach. In strong concoctions of leaves or flowers, or by chewing leaves, the plant causes sweating—done as part of a larger cleansing practice. In some cases, this is said to purify the blood and other parts of the body, and has been used as a treatment for body aches and other maladies. A strong decoction of yarrow can be added to baths to reduce inflammation and increase mobility, especially for the elderly and infirm.

The vapors from heated yarrow are also said to have a role in traditional Quinault medicine. The leaves, flowers, and seed capsules are sometimes heated on an open stove to fill a room with a pleasant smelling vapor, soothing the respiratory tract—a popular remedy for small children who have colds. Heated yarrow may also be used as a general air freshener or as part of the cleansing of people, houses, and objects.

What It Looks Like

Yarrow is a lanky herbaceous perennial with single stems, branching near the top and bearing fine, feathery leaves. Compact umbrella-like clusters of tiny white flowers sit at the top of each stem.

Where to Find It

Yarrow is very common in grassy areas, on roadsides, cleared lots, prairies, and beachfront clearings.

When to Gather

Spring is a common time for gathering yarrow foliage and flowers, when they are fresh and full of active compounds, though the plant is still visible and usable through summer and fall. Fall through spring are common times to gather the roots.

Traditional Management and Care

Yarrow generally grows in clearings and tends to thrive in places where there has been burning or other disturbances that keep the forest at bay. It formerly thrived in places such as prairies, where Quinaults actively managed the land with fire. Patches of plants used for medicinal purposes are often shown respect and sometimes visited repeatedly over the years.

Cautions

Yarrow contains thujone and other active compounds that can be unhealthy for certain people and should especially be avoided during pregnancy.

Dense clusters of tiny white flowers stand on stalks linked with lacy "squirrel tail" leaves on this especially important and widespread medicinal plant.

Fireweed

xwajaʔctont | *Chamerion angustifolium*

Traditional Uses

Fireweed was long an important source of fiber. The cottony material taken from the seed pods of fireweed is traditionally spun into thin cords, serving as a light- to medium-duty string. This cottony substance was also mixed with duck down or other soft materials and spun into a loose fiber that was woven into traditional blankets. The stems could serve as a source of cordage as well, and was used as fishing line and as an all-purpose twine.

Peeling away the outer stem of young shoots exposes a soft, sweet core. Raw or steamed, this sweet interior stem has been used in traditional Quinault foods and desserts. Some families have adapted the plant for use in modern foods, chopping the stem for use in soups and other recipes.

Fireweed stems, leaves, or flowers have also been used in traditional medicine. Simmered, a strong fireweed decoction is added to baths to reduce inflammation and increase mobility, especially for the elderly and infirm. Some sources suggest that fireweed decoctions are not safe to drink, but small quantities are traditionally consumed for respiratory ailments such as coughs, flu, and tuberculosis. Often, these decoctions are made from the root of the plant. Poultices of fireweed have been used on the skin for a range of medical purposes.

The flowers of fireweed, fluorescent magenta to light pink, feature four symmetrical petals and pink stamens dangling from pinkish-white stalks.

What It Looks Like

The tall, single stem of fireweed carries many lance-shaped leaves and terminates in a long spire of attractive magenta flowers. As summer turns to fall, these flowers transform into long seed pods, popping open to reveal seeds within a tangle of cottony fibers.

Where to Find It

Fireweed is widely found along roadsides, meadows, and on recent clear-cuts. Some elders' accounts suggest that

Fireweed grows in open shoreline clearings and, in more recent times, turn forest clear-cuts pink; in both settings, the plant is still harvested for use as food, medicine, and fiber.

the plants used for medicine are best found in shaded settings where the plant struggles.

When to Gather

The edible stalk of fireweed is especially good in young plants that appear in late spring through summer. The cottony fibers can be gathered as the seed pods begin to open on their own—especially in late summer and early fall. The stems can be harvested for cordage in the summer too—when they are no longer springtime-soft nor late-season brittle. The strength of the plant for medicine seems to vary over the year, but little is known; roots are probably more potent in spring, just as the plant emerges or in fall, as the plant goes dormant.

Traditional Management and Care

The growth of fireweed was likely enhanced, like other species, through the use of fire historically. Elders recall that, in places where fireweed was scarce, people took only part of the plant, leaving the rest intact so that harvested plants would continue to live and reproduce. The plants grow in patches from spreading rhizomes, so people can take, even transplant, some plants from a patch and leave others; the remaining population will survive and thrive.

The fibers that emerge from fireweed seed pods in late summer are cottony soft, and are traditionally wound into cords, blankets, and other traditional goods.

Cow Parsnip

waka?	| *Heracleum maximum*

waka? — *Heracleum maximum*

Traditional Uses

Cow parsnip is a large plant known to most Quinault people, seen along roads, riverbanks, and other clearings throughout Quinault country. Sometimes called "Indian celery," this plant has edible stalks of both buds and leaves that resemble celery in flavor and texture. The edible stalks are harvested when the plant is still young, pliable, and sprouting from the ground, and before the flower buds have expanded from their protective sheath. The leaf stalks are cut between the ground and the leaf, while the hollow flowering stalks are cut between the thick "joints" connecting them. Both must be carefully peeled before they are eaten.

Traditionally, cow parsnip is eaten raw or steamed, and dipped in seal oil or something similar. In recent times, Quinault have experimented with new recipes using cow parsnip, from baked goods to stir-fries. Elders report the plant is not only an important late spring vegetable, but also a healthful tonic food, energizing and full of vitamins, as much a "medicine" as a food. Ashes from the charred stalk have been reported as a seasoning and salt substitute for some Northwest tribes.

A poultice of cow parsnip leaves is traditionally warmed and placed on sore joints and limbs. One Quinault term for cow parsnip, *waka?,* literally means "kills the pain," relating to this use. Some sources also mention the use of cow parsnip poultices on sore or swollen eyes. But exercise extreme caution before using cow parsnip for these medicinal applications, as the juices from cow parsnip can cause severe blistering in some people, especially in the presence of bright sunlight.

What It Looks Like

Cow parsnip is a lanky plant with compound, three-parted, maple-like leaves and broad flower heads on tall, fibrous hollow stems. Flower clusters are umbrella-like, with several small white flowers clustered at the ends of the "spokes." Leaves and stalks are covered in fine, light-colored hairs.

"There are these green stalks. Lot of them along . . . the road. They grow in [wet areas]. And they grow in sections—they're round and hollow. They taste something like a celery. We'd be just going along, walking down the trail to the river here, we all were, and we'd just pull one out and peel it and eat it."—Katherine Barr

Where to Find It

Some elders say that cow parsnip "grows almost everywhere!" Especially common in moist clearings along rivers and streams, roadsides, prairies, and other open settings, especially where the forest is open but the ground has not been recently disturbed. Elders note that, while this plant is common along the roadsides, it should not be gathered there because of roadside pollution.

When to Gather

The edible stalk of cow parsnip is especially good in younger plants as they first appear in late spring and very early summer, before flower heads appear. As the summer wears on, stalks become hard and inedible. If the stalks seem too tough and stringy to enjoy, the plant is probably too mature, or the wrong part is being used. So too, the leaves are best used for medicinal purposes when they are young and fresh (see "Cautions").

Traditional Management and Care

Cow parsnip may have been enhanced historically through the use of fire, along with other species. Plants used for medicinal purposes are shown appropriate respects.

Cautions

The juices from cow parsnip stalks and foliage can cause blistering and long-term discoloration of the skin, especially around the lips and face, if it comes into contact with the skin and then the skin is exposed to sunlight. In some people, this effect can be severe. Use cow parsnip with caution, consider using gloves during harvest, only eating stalks after they are peeled and rinsed, and only using medicinal poultices with the advice of a knowledgeable medical professional.

Stalks and leaves of cow parsnip are tender before the flowers begin emerging in the spring and are seldom harvested once flowers appear.

Skunk Cabbage

t̓ogwak̓e | *Lysichiton americanus*

Traditional Uses

A common plant in the wetlands of Quinault country, with its distinctive funky odor, skunk cabbage has a remarkably diverse range of traditional uses. Skunk cabbage is edible—its roots, white and soft in the middle, are peeled and eaten. One Quinault word for the plant, tsulélos, means "digging the roots." This root (actually an underground stem, or "rhizome") can be roasted on rocks, steamed, or cautiously eaten raw. Very hot and peppery, the root is sometimes used, more as a spice than as a food, to season meats and other food items. The fresh spring leaf shoots are also edible, though not usually eaten by Quinaults, while mature leaves are intolerably "hot" and unsafe to eat. The plant, including its root, is also said to have laxative properties, so that some elders view the food as mostly medicinal. Rinsing the roots or shoots repeatedly and cooking them, then changing the water and reboiling, can make the plant more palatable and reduce its laxative effect.

Although a low-priority food plant, skunk cabbage has nonetheless been of short-term importance to the traditional diet. The plant has been described as a famine food in the oral traditions of Chinooks and others—used, for example, when fish runs did not return. Related to this are Chinookan oral traditions describing the man who first called salmon to shore on the Columbia River being honored with an elk-skin blanket and war club before he took the form of skunk cabbage—the blanket and club becoming the distinctively large flower spike and yellow sheath of the skunk cabbage flower. Elk, bear, and deer also eat skunk cabbage, especially in lean times when other browse is limited, and elders note the meat from these animals takes on a skunk-cabbage flavor.

The plant has had an important role in Quinault traditional medicine, and a number of these uses persist to this day. Quinault elders have made a tea from skunk cabbage root and possibly other parts of the plant, using this as a potent laxative for cleansing. This decoction is also said to clean out the bladder and was

"A lot of people don't know much about the skunk cabbage. If you reach down past your elbow, and get down to that part of the root that's just a straight root that grows down, with the little tiny hair-like [rootlets] that go out from it, that one that goes down is all white. If you pull it out and peel all the outside of the root part off . . . it's kind of peppery. It was edible, very edible."
—Chris Morganroth

once popular for this purpose. The plant has been reported as an ingredient in a diverse range of medicines within other tribes: in stimulants and antispasmodics, and as an emetic for certain lung diseases, as well as being used in salves for ringworm, inflammation, and rheumatism. Quinault elders note that a poultice made from the plant, applied directly to the body, is effective for arthritis, joint injuries, and sore muscles. The more severe the injury, generally, the longer the poultice is applied. Some recall seeing elders make a "boot" out of skunk cabbage leaves, wrapped in wax or other materials, to hold against the foot for severe injuries. Some report seeing broken bones or severe joint injuries improved significantly by this treatment.

Large, mature skunk cabbage leaves traditionally serve as "wax paper," too, and are popular in food preparation. Mashed berries can be spread on skunk cabbage leaves and dried in the sun or near a fire, for example, then peeled off to make fruit leather. Skunk cabbage leaves were also used to wrap certain foods such as meat, fish, clover roots, elderberries, or hemlock cambium for pit roasting—a practice that lends spicy flavor to otherwise bland foods.

A Quinault woman gathers skunk cabbage roots, circa 1912.

Skunk cabbage root had other values as well. The thick upper portion of the skunk cabbage root, somewhat round and easily whittled into a small ball, was used in various traditional games. One game, for example, involved tying a root ball to a sharp stick, tossing the root in the air, and repeatedly trying to impale it on the stick as it fell—scoring points as you go. Another game involved making spinning "tops" by piercing a root ball with salmonberry canes. Competitors spun several of these tops simultaneously on a hide, pulled tight over a wooden hoop. The tops that fell were given to one's opponent until the winner had captured all their opponent's tops. The use of skunk cabbage roots in the dyeing of basket materials has also been mentioned—usually to produce dark, near-black colors.

What It Looks Like

The huge, stout, shiny oval leaves of skunk cabbage grow from a common base. Each plant produces a large, yellow flower that is strongly scented, consisting of a single concave bright yellow "sheath" around a prominent, club-like flower spike.

Where to Find It

Skunk cabbage is found exclusively in freshwater wetlands, often in places with shallow slow-moving or standing water during the winter and spring. Lake margins, shallow marshes, and seasonal overflow areas along rivers and streams are especially productive. It is one of the first flowers to make an appearance in the spring.

When to Gather

For food purposes, the early shoots of skunk cabbage are best gathered in the late winter or early spring just as they appear. Roots can be dug at many times, but fall and winter would be typical. The leaves are gathered when they fully mature, in late spring through summer and early fall.

Traditional Management and Care

Those who gather skunk cabbage often only take a portion of its leaves or a fragment of its rhizome, so that the plant will live. Plants are often harvested only selectively from larger patches.

Cautions

The "hot, spicy" effect on the tongue and mouth of skunk cabbage—so popular with some Quinault harvesters—comes from microscopic but sharp calcium oxalate crystals in its tissues. These can become embedded in the mouth and throat, or cause digestive trouble, especially if inappropriate parts of the plant are consumed, or if appropriate parts are consumed in large quantities. This, plus its laxative properties, means that using of skunk cabbage as a food should be undertaken cautiously. If using for food purposes, stick to the root and start with very small samples. Cook and rinse the roots repeatedly. If the plant is unpleasantly hot, it should be avoided. Consult knowledgeable health-care professionals for guidance.

XXII, 4.
40. Urticaceae.
2
1
3
A
WM
6
5
4
7
178. Urtica dioica L.
Große Brennessel.

Stinging Nettle

k̓wənən | *Urtica dioica*

Traditional Uses

In spite of nettles' unpleasant sting, they are among the most valuable of plants in Quinault country, used for traditional food, medicines, and materials. The nettle was especially important in Quinault hunting and fishing technology. This is because the strong fiber of nettle stems can be twined for use in cordage—the source of almost all traditional nets, snares, and rope. Nettle fiber was a primary source of material for fishing nets of many kinds, such as drift nets and dipnets. It was also used to manufacture the nets the Quinault spread above rivers and other low flyways to catch waterfowl. Nettle cord was also used for snares. Quinault hunters would place several nettle twine leg snares amid a flock of bird decoys. Fishing and ocean hunting harpoons often had detachable points secured to harpoon shafts with strong nettle cord of up to thirty feet in length. Finally, nettle rope was sometimes used to tie up and secure Quinault canoes.

The harvest of nettle fiber is done carefully to avoid the stinging spines, often using gloves and targeting the tallest stalks at their full maturity in late summer and fall. Some harvesters formerly made nettle "combs" to run through the stalks to strip the leaves, flowers, and fruiting clusters and to eliminate some of the stinging spines. Some soaked and pounded nettle stalks to loosen their inner fiber, while others dried them in the sun or on indoor mats until they could be pounded to release the fiber lying just below the skin. To make cord, Quinault craftspeople twisted several fibers together to make string—additional fibers being added to increase its thickness. This was sometimes rolled by hand on the thigh, and at times spun with a spindle whorl. By working the loose ends of one strand into the fibers of the next, Quinault people could splice pieces together, making the twine as long as needed. In

"The stalk on the nettles, they would wait until the nettles grew to a certain height. When the flower would begin to flower out and burst into its seeds, that was probably one of the best times to cut it off to make twine. The way my grandmother made the twine was a little bit different from the way other people made it. She would cut off the nettle stalk. Maybe about three or four feet of it.... She'd go just above her knees in the water and she had a little log probably about two feet long.... She would put her nettles into the water and kind of get them soaked, and then beat them on that little log. Take them out and wash off the fleshy parts until she had the green strands that were inside. She would do it until they were nice and soft. Then she'd [lay them out together] one of them straight, next one would be straight, next one would be straight. She wanted to use four of those bunches together.... She'd hold [these strands] tight, down on her lap. And then she'd roll it... so all of those would roll at the same time and it would form a rolled-up piece of twine. It was beautiful, the way she'd do that. It was turning into a beautiful twine that was kind of a light green.... After she got so much turned around into the fibers, she would add some more to it and roll that together. Roll that into it. She could keep doing that until she got it as long as she wanted it. They could go forever. They would make ropes for tying onto fish hooks."
—Chris Morganroth

turn, the strings could be tied together to form nets and other items, traditionally using a net spacer to place the knots and ensure a uniform width of mesh. The nets were soaked and sometimes simmered with tree bark, such as alder or hemlock, to stain the fibers, making the nets less visible to fish or birds. Individual cords were also wrapped together to make rope—usually made of two or four cords wound together.

Nettle is also a food plant and one of the more important spring greens. The new leafy shoots emerging from the ground are picked with care, rinsed in water, then steamed or boiled. That done, the stinging spikes are deactivated and nettle becomes a pleasant, flavorful, and nutrient-rich green—perfect for people emerging from wintertime inactivity and preparing for the busy months ahead.

Nettle also has many medicinal uses. In addition to being consumed as fresh greens, the freshly sprouted leaves are gathered for an early winter and spring tea. The bark of older plants, meanwhile, can be simmered to produce a decoction used for headaches and nosebleeds. The nutrient-rich sprouts were also chewed by pregnant women during labor. As Harvest Moon explains, this was said to "scare the baby out." The long, fully armed stalks were used to "whip" people with certain conditions: the stinging spines helping activate nerves in certain parts of the body when people had suffered paralysis, pinched nerves, arthritis, rheumatism, and similar conditions. Quinault men also sometimes used nettle to whip themselves in order to stay awake and focused while out hunting seals or doing other arduous tasks. In addition, whipping with nettle was sometimes done as part of purification rituals for those who had handled human remains. (Quinault kids have also played intense games of "war" with nettles, chasing and slapping each other with the stinging stalks.) Not surprisingly, as in many Northwestern Native languages, the Quinault term for nettle, *qwunen*, roughly means "it will sting you." Simmered, poultices of nettle are sometimes placed on joints for arthritis and other conditions.

What It Looks Like

Nettle is a tall single-stemmed perennial, often growing in large patches from branching rhizomes, with broad, pointed, saw-toothed leaves lining the stem's sides. Both stalks and leaves are armed with stinging hairs. Small greenish flowers, clustered at the leaf nodes on the upper part of the stalk, droop near the top of the plant.

Where to Find It

Nettle can be found growing in patches on moist, rich soil in forest clearings near marshes, riverbanks and streams, and disturbed areas from low to middle elevations.

When to Gather

For use as fiber, summer and fall is the time of harvest, when nettle stalks are tall and firm but not brittle, and flowers give way to seeds. Greens are usually gathered in late winter and spring for food and medicinal use, but a few patches produce edible new sprouts into the summer, especially if they are trimmed back.

Traditional Management and Care

In some Northwest tribes, people poured fish waste on nettles near fish processing stations to enhance the output of nettle. Fire may have also enhanced nettle patches in some locations by removing dead woody stalks and competing vegetation. Nettle often grows at former village sites, arriving there through ancestors' intentional or unintentional seeding. Cutting back older nettle stalks can sometimes prompt a plant to sprout new young growth, which may have been practiced to combine stalk harvests for cordage with late-season nettle sprout harvests. Elders note that overharvest can cause some nettle patches to entirely disappear—something that has occurred occasionally in Quinault country.

"They always use the nettles for arthritis, too. Slap the hands with it, get it numb then the pain will go away. Just in springtime, though."
—Dennis Martin

Stinging nettle stalks grow tall, lanky, and a bit menacing by summertime, making them ideal for twine-making and related crafts, but beyond their prime for use as food and many medicinal uses.

Cautions

The sting of stinging nettle is unpleasant but temporary. Still, harvesters should use gloves and take other precautions. Those applying nettle directly to the skin for medicinal purposes should be prepared for some interesting discomfort. Mature, tall nettle stalks and leaves—most of what is available in the summer—are inedible. Food use of nettles should involve only the youngest leaves and shoots; the same is true with teas and other internal uses of nettle, unless otherwise advised by a knowledgeable expert.

Palmate Coltsfoot

k̓wayʔax̣ | *Petasites frigidus* var. *palmatus*

Traditional Uses

Palmate coltsfoot, a common plant in the soggier forestlands of Quinault country, is another of the tribe's important medicinal plants. This plant is especially noted as a remedy for symptoms of arthritis, helping reduce pain and increase the mobility of people with that condition. Specific treatments vary slightly, but all involve soaking in an infusion made from coltsfoot. Some elders run a bath, placing the leaves directly in the hot water and soaking in the resulting brew. A few make a strong decoction from simmering the leaves, then pour this into their bathwater for a soak. Additionally, elders often rub themselves with coltsfoot leaves while soaking—especially focusing on areas that are stiff and sore from arthritis. Each approach seems individualized, and appropriate for different sets of symptoms. Short baths are suggested, as the baths can make people lightheaded; and many brief baths with coltsfoot may be more helpful than long soaks. Beyond this use, Quinaults report using coltsfoot for other types of inflammations and infections. Soaks or poultices have been mentioned as a possible aid for nerve problems like diabetic neuropathy. Coltsfoot is sometimes added in small quantities to herbal teas, especially for colds or flus that involve inflammation in the lungs and airways. A mild decoction of coltsfoot has also been suggested as a wash for sore, swollen eyes.

This plant also has some food value. The new spring greens and shoots can be eaten, ideally steamed, but this is not a widespread Quinault practice. Quinault people have sometimes used the leaves to wrap berries when steaming food in pits—perhaps imparting flavor or some medicinal value from the leaves.

Flower stalks of palmate coltsfoot appear early on the forest floor, showing harvesters where to return when seeking fresh green springtime leaves.

"This heals sores . . . but you can't bathe in it for more than fifteen minutes because it's really toxic: it'll make you faint. [One time] my knees and my ankles were all swelled up. . . . My mother put me in the tub—she had to help me get in the tub and out of it—and she had boiled these in it. Just dump everything in the bathtub, the leaves and everything."
—Lucille Quilt

What It Looks Like

Leaves of palmate coltsfoot are long-stalked, with blades that are deeply toothed and mitt-shaped, smooth on top and wooly underneath. Tiny pale flowers cluster like small thistle blooms, appearing atop thick stalks even before the leaves emerge each spring.

Where to Find It

Coltsfoot is found growing in patches on moist ground, in shady woods, seeps, and open wetlands from low to middle elevation. It is also common along roadside ditches, but best avoided in these settings. Plants obtained higher up in coastal mountain ranges are sometimes reported to be more potent than those gathered at lower elevations.

When to Gather

Coltsfoot leaves appear in the early spring and are especially potent at that time. They can be used through the summer and fall where found, but harden with time.

Traditional Management and Care

Burning near marshes and other wetlands may have opened the forest canopy enough to allow coltsfoot patches to expand. As with other medicinal plants, certain respects are shown to plants that are harvested, and harvesters often take leaves selectively to avoid killing entire plants.

Cautions

Palmate coltsfoot contains alkaloids that can be toxic in large quantities or harm the liver; food and tea made from the plant should be used in moderation, and maybe not at all for those with liver complaints. Elders' accounts suggest that the baths can cause dizziness and other complications if they last too long. Starting with small baths might be best for those who are new to coltsfoot, working up to an appropriate level of tolerance.

Palmate coltsfoot often forms dense patches in damp, dimly lit places under the forest canopy, where it is sought for medicinal use—the leaves growing gradually firmer and less potent through the season.

XII, 3.
105. Rosaceae.
4. Potentilleae.
1
2
3
4
5
6
A
408.
Fragaria vesca L.
Gemeine Erdbeere.

Native Strawberries

c̓x̣eʔx̣eʔew (beach strawberry)
x̣eyəx̣ʔim (forest and Virginia strawberry)

Fragaria chiloensis (beach strawberry)
F. vesca (forest strawberry)
F. virginiana (Virginia strawberry)

Traditional Uses

In flavor and appearance, the tiny berries of native strawberries are like miniature versions of their larger, store-bought relatives. Yet they are often better: sweeter, tastier, and more traditional than those bought in stores. Small but often productive, strawberry patches are visited each year around June and July along beaches, on riverbanks, and in grassy meadows throughout Quinault country. A small but tasty harvest, strawberries are often eaten fresh, almost immediately after picking. A few elders have canned strawberries, or made them into jams—purely of strawberries if they have enough, or mixed with other berries if the harvest is small. These sweet berries are traditionally shared with honored guests during feasts and other social gatherings, and young women fanned out to pick them in large quantities for these events.

In addition to discussing their favorite strawberry patches, elders have shared other types of traditional knowledge regarding the berry. Strawberries that grow near the rocky spray zone of the ocean are often unpleasantly salty, for example, though this can be offset by a good repeat rinsing with fresh water. The arrival of blueback salmon (*Oncorhynchus nerka*) on the Quinault and other rivers was said to correspond with the arrival of red and ready-to-eat native strawberries—if one was seen, the other was sure to immediately follow. When non-Indians arrived in the Northwest, some families earned extra money by gathering and selling native strawberries to newcomers, though this work was labor-intensive. Over time, many tribal members applied their berry-picking knowledge at commercial patches of domestic strawberries, finding seasonal work at berry farms through the late nineteenth and twentieth centuries.

In addition to being a healthy berry for food use, the plant has other medicinal values. The leaves, for example,

"[At] the rocks down there at the cove, my kids always went up there and picked strawberries at the old coast guard station. Lot of wild strawberries.... You ever pick a five-gallon bucket full of wild strawberries? Well, I used to go out and get them. My mom made jam and stuff out of them and everything—and you don't dare eat one while you're picking them. If you do eat one, you won't get any home. And as a young kid, do you know how hard it was to pick wild strawberries and not start eating them?! But I'd start thinking about what my mom did with them and that's how I could get them home."—Gerald Ellis

Tiny but delicious wild strawberries are often found hidden on the forest floor.

can be chewed and placed on the skin to soothe and heal burns. The whole plant can be simmered to make a medicine for diarrhea and other digestive complaints, and the leaf was an ingredient in other medicines for a range of maladies, requiring multiple plant products.

What They Look Like

Strawberries have dark, shiny or dull green, three-parted leaves with oval toothed leaflets that grow close to the ground. The plant spreads by trailing runners, growing laterally across the ground. It has simple white flowers—five-petaled and wild-rose-like—that turn into small, sweet red strawberries. Beach strawberry has shiny leaves and rounded berries borne close to the ground; Virginia strawberry is similar but with dull bluish leaves; forest strawberry has elongated berries on taller stalks.

Where to Find Them

Strawberries thrive in exposed open settings, often in rocky places: beach strawberry (*Fragaria chiloensis*) especially grows on the beaches and grassy exposed areas just above along much of the Olympic coast. The other two native strawberries (the "Virginia" strawberry, *F. virginiana*; and the "forest" strawberry, *F. vesca*) grow especially on meadows and prairies and on forest margins. All three can be found on riverbanks and also on older, dry channels up the Quinault and other rivers and streams. Strawberries are occasionally gathered in the mountains, in wet meadows and along stream and riverbanks.

When to Gather

Strawberries appear in late spring and early summer, ripening especially fast in exposed areas such as beaches with direct sun. Some elders report that the berries are available when the bluebacks start to run up the Quinault and other rivers. Fresh springtime leaves are sometimes sought for medicinal use.

Traditional Management and Care

Some elders note that Quinault traditionally fertilized some of their strawberry patches with fish waste or other materials. As Chris Morganroth notes, "Put some fertilizer in there and they'll produce four or five times more berries than natural." Pickers usually try to not trample the small plants and often leave a few berries behind. Some elders note that places where the berries are overharvested and trampled do not come back—at least not quickly. In some settings, such as on prairies and meadows, fire appears to have produced the clearings that allowed strawberry patches to thrive and expand.

Cautions

Strawberries are one of the berries traditionally avoided by pregnant women, as it was said they would cause infants to develop birthmarks. This may relate to the fact that the plant is reported to have some effect on women's reproductive cycles; pregnant women may wish to consult with a knowledgeable physician before consuming teas or medicines made with strawberry.

On beaches and other bare ground, sun-loving "beach strawberries" form large, diffuse patches, their runners expanding each patch step-by-step across bare ground.

Tab. XXVIII.

Vaccinium oxycoccos L.

Wild Cranberry

ʔesuʔumiš | *Vaccinium oxycoccos*

Traditional Uses

The tiny, bright red berries of wild cranberry have long been a welcome, supplementary part of the Quinault diet. Many prefer their sweet-tart cranberry flavor to that of store-bought varieties. People have long picked these berries in the boggier parts of Quinault country, especially in the wet, traditionally burned prairies, where the berries form low mats alongside Indian tea and other wet marsh species. In fact, these berries are so tied to the prairies of Quinault country that one Quinault name for them is asolmix or "prairie berries." While these berries can be eaten fresh from the plant or dried, common practice was to gather and store large quantities in baskets or wooden boxes until they began to soften. In that condition, they could be eaten fresh or wrapped in leaves and pit-cooked with hot stones. Cranberries cooked in this way were a popular autumn food. People still use cranberries in various ways today and experiment with new recipes, such as cranberry preserves or jams.

Long ago, wild cranberries were a trade good within the Pacific Northwest. Quinault oral tradition notes that—with an abundance of the berries in fire-managed wetlands —the ancestors often brought large quantities with them on expeditions to the north and south to trade alongside other plant products. Once traditional management of cranberries ceased in the late nineteenth and early twentieth centuries, the number of berries available in Quinault country declined to a level that surplus harvests were no longer possible, suggesting that traditional management was responsible for these historically high levels of berry production. Elders recall that tribal members formerly got work in commercial cranberry bogs on the southern Washington coast and beyond, employing their knowledge and experience with wild cranberries developed at home on the Quinault prairies.

What It Looks Like

Wild cranberry is a creeping, compact shrubby vine with small leathery leaves widely spaced on finely branched stems. Pink shooting-star-like flowers turn to small tart reddish berries.

"Those [prairies] are pretty neat places. The cranberries! The wild cranberries that grow there . . . those are a good eating berry. And they last a long time on top of the ground, course until it snows, then they start to get soft and turn into mush and then they absorb into the ground."
—Chris Morganroth

Where to Find It

Wild cranberries are found widely in the prairies of the Olympic coast, though patches have been shrinking in recent times. Small patches are found in other peaty wetlands, marshes, lake edges, and similar wetland environments. Experienced harvesters say that different patches have subtly different flavors and that some of the older, well-maintained patches have berries that are larger, with a stronger cranberry flavor.

When to Gather

Cranberries appear in the moist prairies in the summer and persist through the fall—even sweeter after a first frost, but growing mushy and inedible in the winter after multiple frosts or snow.

Traditional Management and Care

Traditional Quinault burning of the prairies helped enhance the size and productivity of cranberry bogs and may have also enhanced the quality of the berries. Without regular burning of the prairies in recent decades, this plant's range has been decreasing. Elders note that these wild cranberry bogs are sensitive—that they can suffer from overharvesting, but also that people have not valued them, allowing grazing, fill, or other interventions to slowly erode Quinault access to these berries. If wild cranberry harvests are to continue in this region, delicate but enduring human intervention will be required.

Wild cranberry often forms sprawling, dense creeping shrubs (above), especially in wetland settings tended carefully over generations; the berries turn bright red when ripe, especially in sunny sites, and their insides become juicy and tart (below).

Wild Mint

kastu (Chehalis term) | *Mentha arvensis*

Traditional Uses

With its brilliantly minty scent and flavor, wild mint is a popular ingredient in medicinal teas. Yet many people drink the tea as a calming, healthy beverage with no specific medicinal purpose. In fact, for some elders, it is a favorite traditional beverage. Mint can be mixed with other ingredients, such as young Douglas-fir needles, to give it a more interesting, complex flavor. Such beverages can be simmered hot or steeped cold for long periods of time to impart the herbal flavor; Quinault elders report that, in darker infusions or large quantities, mint tea serves as a blood thinner and a blood purifier. The tea is also reported to be soothing for cold and flu symptoms. Finally, mint tea, especially a strong infusion or in large quantities, is said to have mild calming effects for anxiety and irritability.

What It Looks Like

Wild mint is an ankle- to nearly waist-high plant consisting of single stems that sprout from a shared, creeping root system; the stems are square in cross section, lined with shiny, bright green, deeply veined leaves with serrated edges and pointy tips. Flowers are tiny and purplish-pink, growing in dense clusters from the leaf nodes along the stalks. Identity can be confirmed by strong minty smell.

Where to Find It

Mint is found especially along river and streambanks, but also lakeshores and the edges of both freshwater and salt marshes. It forms dense patches at places such as Lake Quinault and along the Quinault River, where it is often gathered by travelers and fishermen.

"Another one that they used to gather a lot—still do, as a matter of fact—is mint. Because it's all over along the river bottom. Used a lot of that when we were camping up the river. We used to make tea all the time. . . . It's supposed to have some value to it, too: it calms you down."—Francis Rosander

When to Gather

The leaves of mint, along with leafy stems and flowers, can be gathered as they appear in spring and early summer.

Traditional Management and Care

Harvesters generally avoid overharvesting any one patch or plant, moving between plants along riverbanks and lakeshores. Other forms of care—burning or even fertilization—may have been practiced historically.

Cautions

Some mint relatives, such as pennyroyal, can interfere with pregnancies; pregnant women should consider using extra care—checking with elders or doctors or avoiding wild mints altogether.

Wild mint is recognizable by its potent fragrance but also by its clusters of pale purple to pink flowers growing close to the stalk (above); on streambanks, mint can grow in dense patches, ideal places for a respectful partial harvest (below).

Equisetum Telmateja Ehrh. 764.

Horsetails

teloʔc (common horsetail)	*Equisetum arvense* (common horsetail)
tetoʔc (giant horsetail)	*E. telmateia* (giant horsetail)

Traditional Uses

Seen in almost every bog, low streambank, and ditch of Quinault country, the familiar horsetails can be used for food, medicine, or as a woodworking tool. The shoots of giant horsetail have been widely eaten—specifically, the fertile creamy white shoots that appear before the rest of the plant emerges from the ground. It is because of these shoots that Quinault sometimes have called the shoots, and the entire plant, *tetóts* or *telóts*—literally "to eat it." These stalks are softer and thicker than the rest of the plant and, picked as they first appear in the spring, are sweet, moist, and succulent. More often than not, the shoots are picked, peeled, and eaten raw. The stalks were also occasionally roasted over a fire and steamed, giving them a more tender texture and slightly richer flavor. Sometimes they might even be boiled and mashed like potatoes. Cooked or not, the shoots were often dipped in oil. They are considered to be a spring shoot, eaten at roughly the same time as the sprouts of salmonberry, thimbleberry, and other species. Like those other sprouts, the plant is traditionally seen as healthy and energizing—helping people make the transition from the slow pace and stored foods of winter into the busy warmer months. Though perhaps not as popular as the shoots, the round nodules that appear on giant horsetail roots were also sometimes dug from the ground. Raw or roasted, these were peeled and consumed with whale or seal oil. The black roots also occasionally served as dark design elements in baskets.

The relatively tough green stalks of horsetails contain silicate minerals that the plant draws from the damp soil. This gives horsetails a sandpaper-like roughness that the Quinault and other Northwest tribes put to good use. These "scouring rushes" are used to polish fine woodwork, such as arrow shafts or delicate wood carvings,

"Those horsetail: they're sort of juicy. There's two types of them [and] it's like eating a vegetable." —Phillip Antone Hawks

and to scrape slime off fish. When used as medicine, the stems of horsetail are mixed with willow leaves to make a decoction that is given to girls with irregular periods, helping regulate their menstruation. An infusion made from the roots of giant horsetail served as a wash for sore eyes, sometimes mixed with human milk to enhance the effect. Finally, the juice from horsetail stalks has also been said to serve as a nail polish, being rubbed onto the nails to make them shiny.

When horses arrived in Quinault country, they were so drawn to horsetail plants when browsing for food that some people began calling the plant *moxʷin*, or "horses eat it." In spite of heavy grazing pressures, horsetails still thrive throughout much of the Olympica Peninsula.

"It's the [fertile stalk] of the horsetail. They look like asparagus. . . . When they're just maybe half grown the stalk is sweet and they're always full of water. . . . The water was sweet and probably pure. That was a foodstuff too." —Chris Morganroth

What They Look Like

Common horsetail is a roughly knee-high, green, single-stalked plant with thin, string-like green branches that fan out on all sides at nodule points along the stalk. Collections of plants have a feathery appearance, connected by underground stems or rhizomes. Giant horsetail is similar but larger, growing up to chest height. The edible, spore-bearing reproductive stalk appears before the other stalks appear, with a creamy white color and a spongy, cone-like, spore-bearing head at its tip.

Where to Find Them

Horsetail is found widely in wetlands, riverbanks, and streambanks, and wet ditches throughout the Olympic Peninsula, especially at lower elevations.

When to Gather

The edible whitish stalks of horsetail appear in early spring, before the plant's main green stalks emerge. The root nodules are usually gathered later, especially in the fall.

Traditional Management and Care

Horsetails gathered for medicinal purposes or for use with cultural woodwork are usually shown special respects. Overharvesting is generally not an issue, though some patches have been extirpated by heavy grazing and ground disturbance.

The edible shoots of horsetail are pale and soft, with spongy spore-bearing heads on top.

The tough, inedible leafing stalks are encircled with thread-like green "leaves."

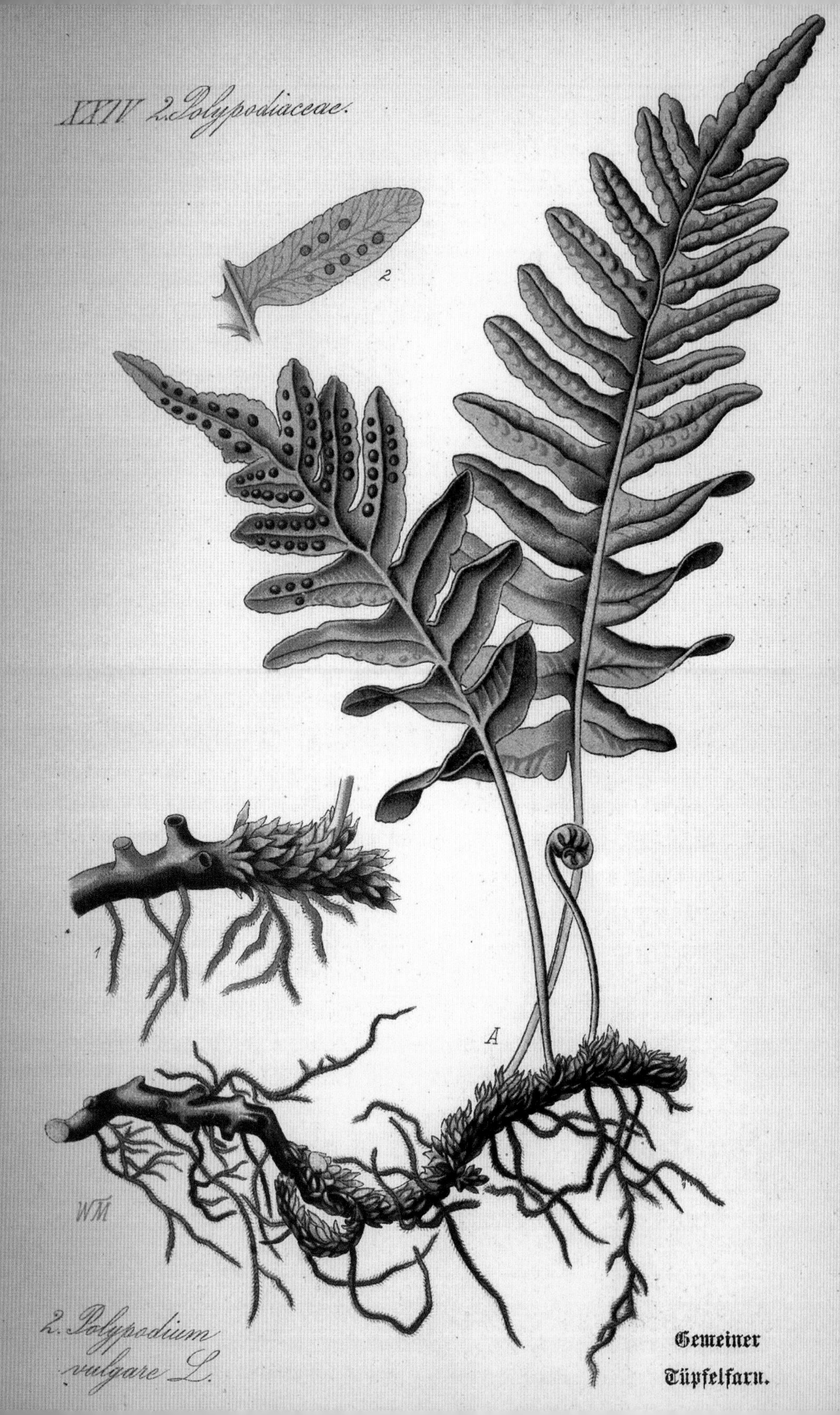
XXIV 2. Polypodiaceae.
2
A
1
WM
2. Polypodium vulgare L.
Gemeiner Tüpfelfarn.

Ferns

p̓la’pla’	Polypodiophyta
sk̓eʔeʔteƛ̓ or čitwašinic (sword fern)	*Polystichum munitum* (sword fern)
c̓umx̣eʔx̣nix̣ (bracken fern)	*Pteridium aquilinum* (bracken fern)
k̓uwa.lsa or cmax̣ays (lady fern)	*Athyrium filix-femina* (lady fern)
k̓u’kutsa (spiny wood fern)	*Dryopteris expansa* (spiny wood fern)
cumanaʔamac (licorice fern)	*Polypodium glycyrrhiza* (licorice fern)
hapalpuɬ (maidenhair fern)	*Adiantum pedatum* (maidenhair fern)
skaʔecƛ̓o (deer fern)	*Blechnum spicant* (deer fern)

Traditional Uses

Ferns are almost ubiquitous in Quinault country; they are the dominant understory plant in many forests, common plants in wetland prairies, and are even found growing from the sides of trees. Quinault people traditionally make use of every species of fern—for food, material, and medicines.

Several ferns have edible "roots" (rootstalks and rhizomes), and these were staple foods for the ancestral Quinault. In fact, some say that fern roots were second only to camas in their importance as plant foods in the traditional Quinault diet. Especially important were the underground sections of bracken, lady, and spiny wood ferns, while sword fern roots would be eaten only in lean times. Individual roots or rhizomes (lateral root-like stems) could be eaten, but in some species—especially wood fern—entire rootstalks, looking like small bunches of dirty bananas, were exhumed, cleaned, and cooked. In addition to these roots, a few Quinault elders describe eating root nodules or "tubers," round growths found on the roots of sword fern, bracken fern, and perhaps other species; eaten raw or roasted, these growths have a flavor and texture reminiscent of chestnut. Women often gathered fern roots at prairies and other inland locations with specialized digging sticks as men fanned out nearby to hunt.

There are many traditional recipes for fern roots. The ferns were often roasted in coals while being rotated, then removed from heat and tapped with a stick or twisted to

Quinault families have used sword fern (*Polystichum munitum*), one of the most common ferns in the Northwest, as food, in food processing, and even in traditional games.

remove the charred scaly outer "skin"—a technique most often applied to bracken fern rhizomes. Fern roots were also steamed in pit ovens for roughly a half day before they were soft, tender, and ready to eat. Like camas, fern roots can be eaten when cooked but have often been mashed, pressed, dried by the fire or in the sun, and stored in baskets for later use. Some fern roots, especially those of bracken fern, were often served with salmon eggs—fresh, dried, or fermented. Others, especially wood fern rootstalks, were traditionally coated or dipped in whale oil when eaten. A "bread" was made by roasting bracken fern rhizomes, pounding or grinding them, mixing the pulverized material with water to make a dough, rolling the dough into small rounds, and cooking these until they resembled small loaves of bread. The Chinooks and others famously brought these loaves to early fur-trading ships at the mouth of the Columbia River.

As is true in many places around the world, in Quinault country the "fiddleheads" from lady and bracken ferns are also important traditional vegetables, gathered as they unfurl in the springtime. These are eaten raw or steamed like asparagus, commonly dipped in whale or seal oil historically. More recently, in addition to eating fiddleheads as a snack or side dish, tribal members have gathered fiddleheads and sold them to commercial buyers for upscale restaurants and markets in places like Seattle.

The mature fronds of many ferns are traditionally used as temporary mats and linings for food processing, including fronds of sword, bracken, and lady ferns. Layered one over the other, these fronds line baskets for food gathering and storage—at once keeping the basket clean and keeping the basket from imparting odd flavors to the food. Layered fern fronds have also been used to make liners for storing food—used, for example, as a lining between layers of clams stored in deep boxes or baskets, keeping each level from sticking to the next. Layered fern fronds are also used to line pits and to wrap food when pit-cooking roots and other foodstuffs. In these situations, people have placed layered fronds over hot rocks within the pit to keep food from burning, and on top of food so the pit can be buried with sand or dirt without food being soiled. The small, leathery fronds of deer fern have also been used to wrap camas bulbs for cooking. By layering sword fern fronds horizontally over the ground, stems aligned at ninety-degree angles to the layer below, people often have made temporary mats on the ground for fish-cleaning and other food processing. Sword ferns have also been popular for wiping slime from fish.

Ferns are also important in the traditional production of fiber and baskets. When bracken ferns mature, the stalks become hard and stringy, and string has sometimes been made from the fibers of the stalks, or from

Found clinging to the sides of mossy trees, licorice fern (*Polypodium glycyrrhiza*) is an important ingredient in traditional medicines.

The dark and durable stems of maidenhair fern (*Adiantum pedatum*), harvested along streambanks, are a principal source of black weaving material, used to produce distinctive designs in traditional baskets.

the underground lateral rhizomes. These rhizomes are sometimes used for the dark weave to decorate baskets. Growing in rocky places near streams and waterfalls, it is the maidenhair fern, though, that is famously linked to Quinault basketry traditions. Maidenhair fern fronds have five or more delicate leafy branches spreading from a central dark black stem. Fiber from the stem is used to produce black designs on baskets made of other more abundant and durable materials. Maidenhair fern stems can be brittle and are usually soaked before being used in baskets; some modern basketmakers soak them with fabric softener or other materials to make them especially strong and flexible.

Ferns such as sword fern are traditionally used in ceremonies and ceremonial regalia. Sword fern is the primary species used in the *pala pala* (literally "fern") game, a traditional Quinault game in which players go up and down the fern frond, touching each leaf and saying *pala!* (fern!) each time, as quickly as possible—a tongue-twister coupled with a test of dexterity.

Growing from rotting logs or soils rich with rotting wood, "wood ferns" (*Dryopteris* ssp.) produce thick root stalks, a traditional food to the Quinault and other tribes.

People customarily look after the ferns. Some harvest these slow-growing plants selectively. Historically, Quinault people intensified the production of bracken fern and other species through the use of fire: in Quinault's prairies, dense bracken fern patches drew deer and elk to graze on the new shoots, allowing Quinault to both hunt and gather ferns at the same places.

Each type of fern has its own distinct medicinal uses

Sword fern was very important in the traditional pharmacopeia: the young leaves of the fern have been chewed and the juice swallowed for sore throat and tonsillitis. Sword fern leaves are also chewed with salmonberry bark to treat boils and sores—the juice is swallowed, but the pulp is sometimes placed on sores as well. Sword fern root nodules are traditionally scraped and used in medicines for headaches, toothaches, and other complaints. The spores from sword fern were used in burn remedies, gathered from the undersides of the leaves and applied directly to the skin.

Bracken fern has been of similar importance to sword fern: fiddleheads are chewed into a paste, spread on wounds, and sometimes covered with banana slug slime to speed healing and avoid infection. Mixed with salmonberry bark, bracken fern was used in poultices for persistent boils and sores. Fiddleheads were also used for toothaches and gum-aches, crushed slightly and held in the mouth against the sore spot.

"Pala pala *is a fern game that the Quinault people played long ago while sitting around the fire in their longhouses; this game is still played today. The game is more [or] less for fun. To play the game you first need to find a long fern [usually sword fern] from the woods. The object of the game is to take turns going up the fern and down touching each petal saying* 'pala' *without hesitation. (Have fun!)"*
—Leilani Chubby

Fronds of *lady fern,* and possibly of other species, have been pounded, boiled, and consumed for labor pains.

Growing on the sides of mossy trees, *licorice fern* has also been a very popular medicinal plant—the long greenish-yellow rhizomes are chewed for sore throats and coughs, or to ease teething pain in infants. This fern is still used by Quinault people to soothe throats after singing at singing circles and other cultural events. It has a semisweet licorice-like flavor, and is used raw or lightly roasted. Licorice fern rhizomes can be chewed for an upset stomach; they also serve as a mild laxative, and elders warn that eating too much can cause diarrhea. Especially among Salish Sea tribes, licorice root has been combined with madrone leaves to make medicines for colds and stomach complaints—a mixture used by some Quinault. Portions of the bracken fern were also chewed as a medicine for chest pain, or bleeding associated with internal injuries. More than one type of fern, bracken among them, was chewed or eaten raw for heart troubles.

"*They used that [bracken fern] to clean the fish with. If they wasn't close to water, that's what they did. They'd butcher the fish, get it ready to dry or smoke. They'd wipe the slime off with the fern. . . . They used the ferns for a lot of things, especially when they're dry.*"
—Katherine Barr

The new, light green fronds of *deer fern* have served in many medicinal preparations, and are traditionally chewed for colic or for lung or stomach complaints. The fronds of this fern are also chewed as an appetite suppressant. Applied directly to the body, deer fern is reported to be used by Quinault elders to reduce inflammation or even improve symptoms of paralysis. A decoction of deer fern is traditionally consumed as a general health tonic.

These ferns also had important cosmetic uses, especially for hair care. Maidenhair fern and lady fern are traditionally charred and the dark ash used as the key ingredient of shampoos to make the hair shiny, dark, and strong. Sword fern leaves were also boiled to make a dandruff shampoo.

What They Look Like

Sword fern is a stalky single-stemmed fern, growing in clumps, with fronds fanning out from a common base. Dark, leathery fronds are commonly 3–4 feet long, but can grow longer. Individual leaflets are largest near the base of the fronds, and the overall frond is feather-shaped, tapering to a pointed tip.

Bracken fern consists of large solitary fronds, standing upright, with foliage spreading, triangular, and multi-branched on the upper half. The stalk is sturdy, with fine orange "hairs," and leaves are feathery.

Lady fern is a fine, lacy-frond fern that is bright green and can reach about 4 feet in height. Frond foliage has a slightly triangular shape and grow in clusters from a common rootstock.

Spiny wood fern is a relatively short, dense fern growing from rootstalks that are usually rooted in rotting logs or stumps. Foliage can look lacy, with serrated leaf tips, and roughly triangular in shape, with the lower part of the stalk clear and without foliage.

Licorice fern is a petite single-frond fern, its fronds growing in patches from gnarled, branched, cord-shaped rhizomes clinging to mossy tree trunks and branches or to rock faces. The fronds are broadly featherlike, with pointed leaflets that are slightly serrated and attached to the central stalk along their base.

Maidenhair fern grows in clusters, with dark-stalked fronds growing up to 20 inches high, each stalk bearing five or more delicate leafy branches spreading out like a fan, giving the frond a lacy appearance.

Deer fern is a small fern with clumped, featherlike fronds growing from a single rootstock, looking somewhat like a small sword fern but reaching no more than roughly 30 inches in height. Simple, single-stem fronds sharing a common base. Each frond is long and narrow, with leathery, dark green leaflets that taper at the base. Mature plants usually produce several upright central spore-bearing fronds, as well as the spreading vegetative fronds.

"*We used to pick up them soft fern on the river bank [lady fern], a chestnut out of that.... It's a fine-leafed fern.... Certain time of year they have chest-nuts on them about like that; really good.*"
—Justine James Sr.

Where to Find Them

Sword fern and deer fern are commonly found growing in the soil of dense forests, including both old-growth and industrial forests, especially at lower elevations. Wood ferns are found in similar settings, but growing primarily on rotting logs and stumps. Licorice fern grows on mossy trees and downed woods in these low-elevation forests—deciduous trees such as bigleaf maple often being the best source.

"We used to get that licorice [fern root] out of the trees: pick it, eat it. . . . It grows out of alder trees and tastes just like that stuff you buy. Only certain areas that I ever seen it. Chew it, eat it. It tastes pretty good. . . . Just break it off and start munching on it."
—Clifford "Soup" Corwin

Bracken fern is found in open woods, moist meadows and roadsides, and on wetland margins from low to subalpine elevations. Bracken fern is still abundant in some traditionally burned prairies. Lady fern usually grows in swampy ground, often in the same general areas as skunk cabbage. Maidenhair fern is very specialized, growing primarily in rocky settings near streams and waterfalls.

When to Gather

Fern greenery can be gathered whenever it is available for use as mats and liners. Greenery used for medicine and basketry is usually gathered when it is still light green and fresh—typically in the summer after it appears.

Times for harvesting edible fern "roots" vary. Bracken fern roots can be dug at many times, but elders report that they are especially good when dug in late summer. Lady fern roots are dug in late autumn. Licorice root gathering often occurs at the same time, but can be done at other times of the year as this plant is needed. Spiny wood fern rootstocks can be dug in fall and throughout the winter and early spring.

Traditional Management and Care

Many ferns are slow-growing and are harvested with great care and selectivity—even exceptionally slow-growing patches of licorice fern and maidenhair fern, for example, can be returned to again and again, so long as one does not overexploit the resource. Those who gather ferns, especially for medicinal or other cultural uses, usually show respects at the time of harvest. The burning of bracken fern prairies was once commonplace; the decline in burning has reduced the density of ferns in those areas.

Cautions

The food and medicinal use of ferns involves some minor hazards. Bracken fern contains potentially carcinogenic compounds and harmful cyanogenic glycosides; concentrations of those compounds vary considerably among bracken patches, and although cooking can reduce these concentrations, it does not eliminate them. In rare cases, the plant can cause hair loss or nutritional deficiencies. While people around the world eat this plant regularly, the health risks have not been clearly established, and so caution and moderation are wise. Some people avoid bracken fern altogether for these reasons, especially uncooked bracken fern, though consumption is still common among certain Native and non-Native harvesters. Other ferns, including wood fern, cause adverse reactions in some people too; although Quinault people have eaten these plants for countless generations, anyone eating these plants for the first time should start with small quantities to be sure that they have no noticeable negative effects and work up from there. There are wood ferns that are not edible; for correct identification of the edible variety, consult a knowledgeable expert. Licorice fern should also be used moderation—elders report that too much can cause diarrhea or, in some cases, high blood pressure.

Image Credits

PLANT PHOTOGRAPHS

Plant photographs in this book are by Douglas Deur, except as noted below.

Trees of the Forest and Forested Riparian Areas

Cascara with berries, mature leaves (2 photos)—Courtesy Jesse Taylor, Wikimedia Commons.
Crabapple—Courtesy Victoria Wyllie de Echeverria, Oxford University.
Cottonwood leaves and bark—Courtesy Schmiebel, Wikimedia Commons.
White pine, cones and needles—Courtesy Richard Sniezko, USFS, Wikimedia Commons.
Vine maple close-up—Courtesy El Grafo, Wikimedia Commons.
Western redcedar, peeled—Courtesy Carla Cole, National Park Service.

Shrubs for Food and Medicine

Blackcap berries—Courtesy Aleazrocha, Wikimedia Commons.
Blackcap blossom—Courtesy Stickpin, Wikimedia Commons.
Bog blueberry—courtesy Algirdas, via Wikimedia Commons.
Oval-leaf blueberry—Courtesy Walter Siegmund, Wikimedia Commons.
Coastal black gooseberry—Courtesy Nancy J. Turner, University of Victoria.
Indian tea plant, leaves—Courtesy Leilani Chubby, Quinault Indian Nation.
Indian tea field—Courtesy Courtesy Larry Workman, Quinault Indian Nation Communications.
Indian tea, flowering—Superior National Forest/Albert Herring, Wikimedia Commons.
Devil's club berries close-up—Courtesy Murray Foubister, Wikimedia Commons.
Oregon grape, berries—Courtesy The Marmot, uploaded by Amanda44, Wikimedia Commons.
Wild rose hip—Courtesy Nancy J. Turner, University of Victoria.
Wild rose—Nootka rose patch—Courtesy Hawkwing3141, Wikimedia Commons.

Other Plants of Meadows, Wetlands, and the Forest Floor

Beargrass meadow—Courtesy National Park Service.

Camas bulb bundle—Courtesy Zeke Serrano, Quinault Indian Nation.

Cattail close-up—Courtesy Dominicus Johannes Bergsma, Wikimedia Commons.

Cranberry close-up—Courtesy Mariluna, Wikimedia Commons.

Cranberry bog photo—Courtesy B. Lezius, Wikimedia Commons.

Fireweed with seed pods opening—Courtesy W. Carter, Wikimedia Commons.

Maidenhair fern—Courtesy Stan Shebs, Wikimedia Commons.

Horsetail—inedible stem close-up—Martin Kozák, via Wikimedia Commons.

Sweetgrass (*Schoenoplectus*)—Rob Routledge, Sault College, Courtesy Wikimedia Commons.

Sweetgrass (*Hierochloe*)—Courtesy Krzysztof Ziarnek, Wikimedia Commons.

Wild mint, flowering (2 images)—Courtesy Leo Michaels.

Wild mint, front image—Courtesy Ivar Leidus, Wikimedia Commons.

HISTORICAL PHOTOGRAPHS AND LINE DRAWINGS

Trees of the Forest and Forested Riparian Areas

Canoe crossing, Quinault River—Edward S. Curtis, *The North American Indian*. Files courtesy University of Washington.

River canoes, Quinault River—Edward S. Curtis, *The North American Indian*. Files courtesy University of Washington.

Hemlock dipnet and fishing weir (sketches)—Ronald Olson, *The Quinault Indians*.

Maple—paddles and dipnet handle (sketches)—Ronald Olson, *The Quinault Indians*.

Vine maple—woman with tumpline basket—Courtesy Leilani Chubby, Quinault Indian Nation.

Yew—tools and bailers (sketches)—Ronald Olson, *The Quinault Indians*.

Shrubs for Food and Medicine

Maggie Kelly picking berries—Edward S. Curtis, *The North American Indian*. Files courtesy University of Washington.

Other Plants of Meadows, Wetlands, and the Forest Floor

Skunk cabbage gatherer—Edward S. Curtis, *The North American Indian*. Files courtesy University of Washington.

OTHER PHOTOGRAPHS

Front Matter

Watching the canoes arrive—Brandi Montreuil—Courtesy *Tulalip News*.

Ethnobotany in the Land of the Quinault.

Arriving on the Quinault beach by canoe—Courtesy Larry Workman, Quinault Indian Nation Communications.

Clam baskets on the beach—Courtesy Larry Workman, Quinault Indian Nation Communications.

Modern prairie burning—Courtesy Larry Workman, Quinault Indian Nation Communications.

Camas digging—Courtesy Zeke Serrano, Quinault Indian Nation.

Sweetgrass harvest, Harvest Moon—Courtesy Quinault Division of Natural Resources.

Trees of the Forest and Riparian

Cedar bark peeling—Courtesy Larry Workman, Quinault Indian Nation Communications.

Peeled cedar bark—Courtesy Larry Workman, Quinault Indian Nation Communications.

Children peeling cedar bark—Courtesy Leilani Chubby, Quinault Indian Nation.

Western redcedar bark processing—Courtesy Quinault Division of Natural Resources.

Shrubs for Food and Medicine

Devil's club gatherers—Courtesy Leilani Chubby, Quinault Indian Nation.

Salmonberry basket—Courtesy Kim Recalma-Clutesi, Qualicum First Nation.

Other Plants of Meadows, Wetlands, and the Forest Floor

Sweetgrass harvester—Courtesy Leilani Chubby, Quinault Indian Nation.

PAINTED BOTANICAL ILLUSTRATIONS

Trees of the Forest and Forested Riparian Areas

Cascara—Franz Eugen Köhler, 1897. *Köhler's Medizinal-Pflanzen.*

Cottonwood—François André-Michaux, 1810. *Histoire des arbres forestiers de l'Amérique septentrionale . . .*

Crabapple—*Curtis's Botanical Magazine*, 1919. Royal Botanic Gardens, Kew; Stanley Smith Horticultural Trust.

Douglas-fir—P. Mouillefert, 1892–1898. *Traité des arbres et arbrissaux, Atlas.*

Red alder—C. A. M. Lindman, 1901–1905, *Bilder ur Nordens Flora.*

Western redcedar, Sitka spruce, bigleaf maple, vine maple—François

André Michaux, Thomas Nuttall, and J. J. Smith, 1865, *The North American Sylva.*
Yew—Otto Wilhelm Thomé, 1885, *Flora von Deutschland, Österreich und der Schweiz.*

Shrubs for Food and Medicine

Blackcaps—E. I. Schutt. Courtesy USDA Agricultural Research Service.
Blackberry—John E. Sowerby, 1863–1877. *English Botany, or Coloured figures of British Plants.* J. Syme, ed.
Devil's club—*Curtis's Botanical Magazine,* 1914. Royal Botanic Gardens, Kew; Stanley Smith Horticultural Trust.
Indian tea—William Miller, 1817–1833, *The Botanical Cabinet.*
Oregon grape—Mary E. Eaton, 1917, "Our State Flowers: The Floral Emblems Chosen by the Commonwealths," *National Geographic Magazine* 31.
Salal—*Curtis's Botanical Magazine* 1828. Royal Botanic Gardens, Kew; Stanley Smith Horticultural Trust.
Salmonberry—E. H. Harriman and C. H. Merriam, 1901, *Harriman Alaska Expedition.*
Thimbleberry—*Edward's Botanical Register,* 1830. Courtesy Botanicus, Missouri Botanical Garden.
Wild Rose—John Lindley, 1821. *The Royal Horticultural Society Diary.* Courtesy BernardM, Wikimedia Commons.

Other Plants of Meadows, Wetlands, and the Forest Floor

Beargrass—Frederick Traugott Pursh, 1846. *Flora Americae Septentrionalis.*
Camas—*Curtis's Botanical Magazine,* 1827. Royal Botanic Gardens, Kew; Stanley Smith Horticultural Trust.
Cattail—C. A. M. Lindman, 1901–1905, *Bilder ur Nordens Flora.*
Fern, Fireweed, stinging nettle, *Schoenoplectus* sweetgrass, wild strawberry—Otto Wilhelm Thomé, 1885, *Flora von Deutschland, Österreich und der Schweiz.*
Horsetail—Christiaan Sepp, in Jan Kops, 1849, *Flora Batava of Afbeelding en Beschrijving van Nederlandsche Gewassen.*
Wild cranberry—Adolphus Ypey, 1813, *Vervolg ob de Avbeeldingen der artseny-gewassen met derzelver Nederduitsche en Latynsche beschryvingen.*
Yarrow—Johann Georg Sturm, 1796, *Deutschlands Flora in Abbildungen.*

MAPS

Map of Olympic Peninsula vegetation zones. Courtesy David Banis, Portland State University.

Selected Glossary

annual—A plant that lives only for a single year.

catkin—A soft, often dangling flower spike on trees such as red alder and willow that serves as a single-sex flower on wind-pollenated trees.

culturally modified tree (CMT)—A live tree that bears the effects of wood, bark, or pitch harvesting, such as western redcedar trees with a strip of bark harvested from the side.

decoction—A tea or liquor that concentrates the essence of a plant by boiling it in water or, less commonly, oil. Decoctions can be made from roots, stems, flowers, leaves, or other plant parts.

emetic—A medicine that intentionally results in vomiting, such as for cleansing the body or expelling poisons.

infusion—A liquid made by soaking plant materials in water or other fluids to extract flavors or medicinally active compounds.

girdling—Removing bark from the circumference of a tree completely, so that the tree dies.

perennial—A plant that lives for more than two years.

pit-cooking—A traditional method of cooking in which a pit is dug in the ground and a fire built in the hole and lined with stones; when the fire has died down, food is placed in the pit, water is usually placed on the rocks to produce steam, and the pit is covered with soil and other materials, allowing food to cook slowly belowground.

purgative—A substance that causes purging, especially through diarrhea or other rapid bowel movements.

rhizome—A rootlike stem, usually growing horizontally above or below the soil, found on such plants as sedges and springbank clover.

stamen—The male part of a flower, consisting of a pollen-producing anther spike on a thin, stem-like filament.

Selected Sources on Quinault Plant Use

Charles, Beatrice, Vince Cooke, Elaine Grinnell, Chris Morganroth III, Lela Mae Morganroth, Melissa Peterson, Viola Riebe, Adeline Smith, and Jacilee Wray. 2004. *"When the Tide Is Out": An Ethnographic Study of Nearshore Use on the Northern Olympic Peninsula.* Port Angeles, WA: Olympic National Park and the Coastal Watershed Institute.

Chubby, Leilani A. 2002. *History and Culture of the Quinault Indians of the Pacific Northwest Coast.* Taholah, WA: Quinault Cultural Center and Museum.

Compton, Brian D. 2003. Western Washington Tribes Ethnobotanical Literature Review. Unpublished report. Olympia, WA: Northwest Indian Fisheries Commission.

Curtis, Edward S. 1913. *The North American Indian, Volume 9: The Salishan Tribes of the Coast.* Norwood, MA: Plimpton Press.

Farrand, Livingston 2002. "Traditions of the Quinault Indians." In *Memoirs of the American Museum of Natural History, Volume IV, Publications of the Jesup North Pacific Expedition,* vol. 3, pp. 77–132. New York: Knickerbocker Press.

Gunther, Erna. 1973 [1945]. *Ethnobotany of Western Washington.* Seattle: University of Washington Press.

Gunther, Erna. n.d. Erna Gunther papers. Unpublished ms. Doc. 591; Accession 1966. University of Washington Libraries, Special Collections, Seattle.

Hajda, Yvonne. 1990. "Southwestern Coast Salish." In *Handbook of North American Indians, Volume 7: Northwest Coast,* pp. 503–517. Washington, DC: Smithsonian Institution.

Harrington, John Peabody. n.d. Ethnographic notes on Quinault/Chehalis/Cowlitz/Yakima/Chinook/Chinook Jargon. Unpublished ethnographic notes. The Papers of John P. Harrington in the Smithsonian Institution 1907–1957. Vol. 1, Reel 17-18. Washington, DC: Smithsonian Institution.

Henry, G. A. 1872–1876. Annual Reports of Quinault Indian Special Agent, G.A. Henry to Commissioner of Indian Affairs (various). Washington, DC: US Government Printing Office.

Hill, Joseph 1865–1868. Annual Reports of Quinault Indian Sub-agent Joseph Hill to Washington Indian Superintendents (various). Washington, DC: US Government Printing Office.

James, Justine E. Jr. 2000–2017. Cultural Field Notes on Quinault Prairies and Ethnobotany. Unpublished field notes, in possession of Justine James Jr., Quinault Indian Nation Department of Natural Resources.

James, Justine E. Jr. n.d. Culturally Significant Native Plants. Table. Taholah: Quinault Indian Nation Department of Natural Resources.

James, Justine E., and Leilani A. Chubby. 2002. "Quinault." In *Native Peoples of the Olympic Peninsula, Who We Are*, edited by Jacilee Wray, pp. 97–117. Norman: University of Oklahoma Press.

James, Karen, and Victor Martino. 1986. Grays Harbor and Native Americans. Unpublished report. US Army Corps of Engineers, Seattle District. Seattle, WA.

Jones, Joan Megan. 1977. *Basketry of the Quinault.* Taholah, WA: Quinault Indian Nation.

Masten, Ann Maria. 2003. Interview on Quinault Traditions. *Nugguam.* Clarinda Underwood, interviewer and editor. Taholah, WA: Quinault Indian Nation.

Moerman, Daniel E. 2010. *Native American Food Plants: An Ethnobotanical Dictionary.* Portland, OR: Timber Press.

Olson, Ronald. 1936 [1967]. "The Quinault Indians." *University of Washington Publications in Anthropology* 6 (1): 1–194.

Olson, Ronald. n.d. Ronald L. Olson Field Notebooks. Unpublished notebooks. Doc. 1047; Accession 1966. University of Washington Libraries, Special Collections, Seattle, WA.

Peter, D., and D. J. Shebitz. 2006. "Historic Anthropogenically-Maintained Beargrass Savannas of the Southeastern Olympic Peninsula." *Restoration Ecology* 14 (4): 605–615.

Powell, Jay, and Chris Morganroth III. 1998. Quileute Use of Trees and Plants, a Quileute Ethnobotany. Unpublished report. La Push, WA: Quileute Nation Natural Resources Department.

Quinault Cultural Center and Museum. n.d. Shoalwater Bay Plants. Leilani Chubby, author. Unpublished ms. in collections of Quinault Cultural Center and Museum, Taholah, WA.

Quinault Indian Nation. 1993. The Quinault Indian Nation Guide to Agroforestry. Unpublished report. Seattle and Taholah, WA: Quinault Indian Nation Department of Natural Resources.

Reagan, Albert B. 1934. "Plants Used by the Hoh and Quileute Indians." In *Transactions of the Kansas Academy of Science*, vol. 37, pp. 55–70. Lawrence: Kansas Academy of Science.
Ryan, Teresa Loa. 2000. Defining Cultural Resources: Science, Law, and Resource Management for Sweetgrass *Schoenoplectus pungens* in Grays Harbor, Washington. Master of Science Resource Management Thesis, Central Washington University.
Seaburg, William R., Bob Pope, and Harry Shale. 2009. "Sun and Moon Are Brothers: A Traditional Quinault Story." In *Salish Myths and Legends: One People's Stories*, edited by M. T. Thompson and S. Egesdal, pp. 202–209. Lincoln: University of Nebraska Press.
Shebitz, Daniela Joy, Sarah Hayden Reichard, and Peter W. Dunwiddie. 2009. "Ecological and Cultural Significance of Burning Beargrass Habitat on the Olympic Peninsula, Washington." *Ecological Restoration* 27 (3): 306–319.
Singh, Ram Raj Prasad. 1966. *Aboriginal Economic System of the Olympic Peninsula Indians, Western Washington.* Sacramento, CA: Sacramento Anthropological Society, Sacramento State College.
Singh, Ram Raj Prasad. n.d. Letters from R. R. P. Singh to Melville Jacobs. University of Washington Libraries, Special collections. Melville Jacobs Collection. Accession 1093-71-130. Box 112 Folder 16.
Stanley, Curt, and QDNR Staff. 1983. An Interview with Horton Capoeman. *QDNR Department of Natural Resources Newsletter*, Fall 1983. Taholah, WA: Quinault Indian Nation DNR.
Stevens, Isaac, et al. 1856. Treaty with the Quinaielt, etc. 1855. (Signed on the Quinault River, WA, July 1, 1855 and in Olympia, WA January 25, 1856; ratified by U. S. Congress March 8, 1859). Reprinted 1904, C. J. Kappler, ed. *Indian Affairs: Laws and Treaties, Volume 2: Treaties*, pp. 719–721. Washington, DC: US Government Printing Office.
Storm, Jacqueline M., D. Chance, J. Harp, K. Harp, L. Lesteele, S. C. Sotomish and L. Workman. 1990. *Land of the Quinault*. Taholah, WA: Quinault Indian Nation.
Storm, Linda, and Daniela Shebitz 2006. "Evaluating the Purpose, Extent, and Ecological Restoration Applications of Indigenous Burning Practices in Southwestern Washington." *Ecological Restoration* 24 (4): 256–268.
Swan, James. 1972 [1857]. *The Northwest Coast; or Three Years' Residence in Washington Territory*. Seattle: University of Washington Press.

Willoughby, Charles Clark. 1886. Indians of the Quinaielt Agency, Washington Territory. *Annual Report of the Board of Regents of the Smithsonian Institution*. Washington, DC: Smithsonian Institution.

Willoughby, Charles Clark. n.d. Unpublished notes, 1883–1888. Charles Willoughby papers. Doc. 1680; Accession 1957. University of Washington Libraries, Special Collections, Seattle.

Wray, Jacilee. 2014. Notes on Ethnobotany of Tribes Associated with Olympic National Park. Unpublished notes. Port Angeles, WA: USDI National Park Service, Olympic National Park.

Wray, Jacilee, and Justine James Jr. n.d. History of the Queets Indians. Unpublished ms. in possession of the Quinault Indian Nation Department of Natural Resources and Olympic National Park, Taholah and Port Angeles, WA.

Wray, Jacilee, M. Kat Anderson, and the Olympic Peninsula Intertribal Cultural Advisory Committee. 2015. Culturally Significant Plants within the Olympic National Park That Have Contemporary Uses for the Eight Olympic Peninsula Tribes. Unpublished notes in the collection of Justine James Jr., Quinault Indian Nation Department of Natural Resources.

Index